PLAYING THE GAME OF DRONES

ALSO BY DR. MELISSA K. AHO

A Short History of the Mosque in Minnesota

The Machiavellian Librarian: Winning Allies, Combating Budget Cuts, and Influencing Stakeholders (co-edited with Erika Bennett)

PLAYING THE GAME OF DRONES

EXAMINING THE ROLE OF U.S. DRONE STRIKES IN U.S. AND ENGLISH-SPEAKING ALLIES NEWSPAPERS

Dr. Melissa K. Aho

SECOND GOLDFISH
Minneapolis, Minnesota

First edition June 2024

SECOND GOLDFISH

Published by Second Goldfish LLC
Minneapolis, Minnesota

A catalog record for this book is available from the Library of Congress.

ISBN 978-1-7342769-2-3 (hardcover)
ISBN 978-1-7342769-3-0 (paperback)
ISBN 978-1-7342769-4-7 (e-book)

Published by Second Goldfish LLC
Contact us at info@secondgoldfish.com
Visit Second Goldfish at secondgoldfish.com

ABSTRACT

In the years following the terrorist attacks on September 11, 2001, the United States ramped up its usage of drones and drone strikes around the world. Spanning three United States presidents, drone strikes became a regular feature in the U.S. military arsenal. While American newspaper media and citizens have been very pro-drone, global citizens view drones in a far more negative light. This study examines U.S. military drone strikes and English-speaking allies' newspapers in Australia, Canada, New Zealand, and the United Kingdom and evaluates if coverage remains positive or negative depending on the newspaper's conservative or liberal leanings from 2008–2019. The book explores how political leanings in U.S. English-speaking allies' newspapers do have some influence if the articles are positive or negative towards U.S. drone strikes and finds that it's not a one size fits all situation. Interestingly, this study also found that as the years of war continued, the number of drone strike articles found in the U.S. and her English-speaking allies' newspapers decreased, perhaps reflecting a donor fatigue situation.

DEDICATION

This book is dedicated to my family Carole Aho, Terry Aho, Michelle Aho, Che Aho, Candy Aho, and Ginger Aho with much love. I also dedicated this book to my late father Terry Aho and my late grandmother Katherine Gulczinski. Grandma would have loved to have a 'Doctor' in the family. She once gave me a graduation card after one of my graduation ceremonies and said in front of my younger cousins, "See, every time you get a degree you get money." If only that were true Grandma, if only.

CONTENTS

LIST OF TABLES

LIST OF ILLUSTRATIONS

LIST OF ABBREVIATIONS

CIA	Central Intelligence Agency
UAV	Unmanned Aerial Vehicle
UK	United Kingdom
US	United States of America

CHAPTER ONE

INTRODUCTION

Drones were not a popular tool in the US war chest before the terrorist attacks on September 11, 2001. They were used here and there but mostly for intelligence gathering by both the US military and the Central Intelligence Agency (CIA). Using drones and drone strikes against an enemy was something that other countries, like Israel, did. Then during the 1990s, the United States used drones in the Yugoslav Wars. They were popular for their "ability to loiter" and for the "use of transmitters to send the footage straight back to battlefield commanders," which was "credited with convincing NATO generals they needed to start bombing the Serbs again, so hastening the signing of the Dayton peace accords" (The Bureau of Investigative Journalism 2019). By 2000 the United States Air Force and CIA put a drone together with a missile as "part of a failed CIA attempt to kill Osama bin Laden" (The Bureau of Investigative Journalism 2019). Then the terrorist attacks on September 11, 2001, happened and the drone's usage changed rather quickly. As the United States' 'War on Terror' continued and expanded, so too did the use of the drone strike, from Pakistan to Afghanistan, Somalia, Iraq, Syria, and Libya, and with it was the rise of death by drone. Drone strikes in the new millennium have become the new normal for US armed forces.

Statement of the Problem

Military drones and drone strikes are now part of the industrial-military complex, but those that support and oppose their usage are not as it seems. After almost eighteen years of American military forces using drones around the world, one could easily assume that our English-speaking allies were equally opposed, as global public opinion is, but that does not seem to be the case. Due to the US government's stance on not releasing drone strike data to the public, most citizens of the United States do not realize that they continue to occur. The lack of drone data released by the US government has left a void in academic literature. Few academic studies have analyzed or questioned drone strikes. This study examines US drone strikes and English-speaking allies' newspapers in Australia, Canada, New Zealand, and the United Kingdom. It argues that while US newspaper support for drone strikes continues to be positive, with our closest English-speaking allies, it conversely continues to decrease.

Research Questions

This book asks the following questions:
Has American newspaper coverage of US drone strikes remained positive from 2008–2019?
Has coverage opposing US drone strikes increased in English-speaking newspapers of US English-speaking allies from 2008–2019?
Do the political leanings of newspapers influence the narrative for support for or opposition to US military drone strikes from 2008–2019?

Significance of the Study

This study of US drone strikes is significant because they are a feature of the American war machine for almost eighteen years, and scholarship on drones is still in the beginning stages. Drone strikes have become increasingly popular, even though the American public has no idea that they are happening. There is a vague popular idea in media from television and the movies of a lone American soldier sitting in a bunker somewhere in the United States and seeing a terrorist on their computer screen half a world away in Afghanistan or Syria and directing a drone to take out that one bad person. While this can happen, what usually happens is that drone strike kills not only that one person but everyone around them, regardless of guilt or innocence. In real life, images can be blurry, people make mistakes, hardware and software breakdown, and that one drone strike can kill innocent men, women, and children indiscriminately.

There is also no trial or jury for those lives and usually no compensation from the US government for killing the wrong individuals. No data has been released on drone strikes by the US government, no photos have been released before, during, and most importantly after the drone has released its missile. No images are ever seen of the deaths and destruction caused by drones or the second drone (called the 'double tap') that often go in to kill for those, such as relatives and medical individuals, that come out to help the survivors.

Key Terms & Definitions

Drone

A drone has "come to refer to 'unmanned aerial vehicles' (UAVs), or what the Federal Aviation Administration (FAA) refers to as an 'unmanned aerial system' (UAS) and the US Air Force to as a 'remotely piloted aircraft' (PRA)" (Kreps 2016, 7). They are best described as "a category [that] refers to any unmanned, remotely piloted flying craft, ranging from something as small as a radio-controlled toy helicopter to the 32,000-pound, $104 million Global Hawk. If it flies and it's controlled by a pilot on the ground, it fits under the everyday-language definition of drone" (Atherton 2013). In other words, a drone is "a machine that performs a preprogrammed task with or without human interaction," and it should be noted that a "true drone does not make independent decisions, although it may appear to do so to the outside observer" (Springer 2013, 2).

Drones are unique in that they are remotely piloted or unmanned. Drones are also meant to be recovered and can loiter or hover over a target, whereas a cruise missile or a landmine are one-way systems and are not meant to return home (Kreps 2016, 8). Boyle also helps define drones as having "two distinct functions: (1) the collection and recording of images and data from the external environment; and (2) the delivery of payloads, ranging from food, medicine, and commercial packages to missiles" (2020, 9). Boyle also defines what drones are not, they "are slower than most forms of manned aircraft and can be less precise than a cruise missile; they are no more automatic or dependent on computers than commercial airlines. Drones can see things on the ground just as satellites can, but not necessarily at a higher level of detail. Drones are less adaptable and nimble than manned aircraft and, at present, many military drones, such as the Reaper or Global Hawk, are less likely than a manned aircraft to evade enemy fire" (2020, 10).

The term drone can be used for ones that are used in the air, ones that are used on the ground, and ones that are used in the water. This book will focus on drones that are used in aviation.

Military Drone Strike

A drone strike occurs when a drone releases a "missile or set of missiles fired at a single location in a short time window" (The Bureau of Investigative Journalism 2019). The United States often launches 'signature strikes' "which target not specific individuals but instead groups engaged in suspicious activities" (Byman 2013). The Israelis first coined the term 'targeted killing,' but Americans prefer the term "'drone strike' to talk about killing terrorists through the use of drones" (Rinehart 2018, xxiv). Other terms that have also been used are "signature strikes, long-ranged hot pursuits, selective targeting, surgical eliminations, decapitation, targeted thwarting, manhunts, active self-defense, pinpoint prevention, and focused disruption" (Rinehart 2018, xxiv). Rinehart (2018) writes that Nils Melzer (2009, 3–6) argues that generally targeted killings have "five elements in common:

1. Targeted killings involve the use of lethal force.
2. Targeted killings are designed to target specifically identified individuals.
3. Targeted killings are carried out with the deliberate intent to kill the individual in question.
4. Targeted killings are used against individuals not in the physical custody of the state.

5. Targeted killings are carried out by states and their central governments" (Rinehart 2018, xxiv-xxv).

Melzer (2009, 3–6) also argues that killings from non-state actors can also qualify under this category.

Finding drone strike data is a complicated process, as the US government does not release this information. Non-governmental organizations, such as the Bureau of Investigative Journals and New America, search out and report this data. They admit that their numbers are a minimum number of drone strikes that are known to have happened, but they rely on a variety of sources and are constantly updating their data as new information comes to light. Table 1 shows the known drone strikes from Pakistan, Afghanistan, Somalia, Yemen, Libya, and Iraq. Question marks are listed for years where strikes are known to have happened, but that information is spotty and no good number has been released, so this author decided to not list this incomplete data, but also felt that the information was valuable for the reader to know that drone strikes did happen. It is also likely drone strikes happen in years where nothing is listed.

Deaths/injuries from US military drone strikes

The actual number of dead and injured from the US military drone strikes is a complicated number to find, as the US government refuses to release those numbers. Rarely does the US government admit that anyone other than militants have been killed or injured. Any male older than a child tends to be labeled a militant, so this broad definition labels almost every male a militant. Pakistani lawyer Shahzad Akbar told the Canadian newspaper *Globe and Mail* that "the whole burden of proof has been reversed by the U.S. in the public narrative: You are killed in a drone strike in North Waziristan, so you are a militant until you come out of your coffin and say otherwise" (Nolen 2012).

The Bureau of Investigative Journalism, a British non-profit news organization that pursues and investigates public interest areas, has been investigating the number of total individuals killed due to US drones strikes and reports that since 2004 out of the 6,000-minimum confirmed known drone strikes, between 8,000–12,000 total individuals, including civilians and children, have been killed (The Bureau of Investigative Journalism 2021). Data from the number of deaths and injuries from US drones comes from the Bureau of Investigative Journals and New America website. The New America is a non-partisan think tank based in Washington DC, which focuses on a range of public policy issues, including drones (New America 2019).

Table 1. Minimum drone strikes as of April 6, 2021, for Pakistan, Afghanistan, Somalia, Yemen, Libya, and Iraq (The Bureau of Investigative Journalism 2021; New America 2021).[1]

Year	Pakistan	Afghanistan	Somalia	Yemen	Libya	Iraq
2001		?				
2002		?		1		
2003		?	?	0		
2004	1	?	?	0		
2005	3	?		0		
2006	2	?		0		
2007	5	?	3	0		
2008	38	?	0	0		?
2009	54	?	1	1		?
2010	128	?	0	2		
2011	75	?	1	13	0	?
2012	50	?	2	41	2	
2013	27	?	1	22	3	
2014	25	?	3	17		
2015	13	235	11	21	2	
2016	3	1071	14	37	513	
2017	5	2609	35	127	17	
2018	1	1985	45	36	6	
2019		7167	63	8	7	

[1] United States drones have been striking Afghanistan since late 2001, but the data is spotty at best. Between 2008-2012 in Afghanistan, numbers include the United States, the United Kingdom, and allies forces drone strikes, and after January 1, 2015 only the United States was launching drone strikes (The Bureau of Investigative Journalism 2021). United States drone data for Iraq and other countries, is spotty and difficult to find as it is published by independent organizations that verify after the fact, as drone strike data is not published by the United States government. The numbers change as new evidence on drone strikes are found and verified by various global NGOs and this author has picked the minimum numbers for both Table 1 and Table 2, which keep estimates low.

Table 2. Minimum killed by drone strikes as of April 6, 2021, for Pakistan, Afghanistan, Somalia, Yemen, Libya, and Iraq (The Bureau of Investigative Journalism 2021; New America 2021).

Year	Pakistan	Afghanistan	Somalia	Yemen	Libya	Iraq
2001						
2002				6		
2003				0		
2004	6			0		
2005	16			0		
2006	94			0		
2007	36		21	0		
2008	252		0	0		
2009	471		2	30		
2010	755		0	5		
2011	362		2	76		
2012	212		4	217		
2013	109		2	79	4	
2014	115		10	90		
2015	60	982	20	75	9	
2016	11	1388	204	153	73	
2017	15	876	216	135	121	
2018	1	778	335	31	19	
2019		118	326	1	43	

Both the Bureau of Investigative Journals and New America tend to underestimate the number of drone strikes (Table 1) and deaths (Table 2), as they rely on a variety of sources, and are constantly updating their data as new information comes to light. Table 2 illustrates the minimum number of drone strike kills in this study from Pakistan, Afghanistan, Somalia, Yemen, Libya, and Iraq. It is interesting to note that most deaths that took place in Afghanistan and Somalia happened in the latter years of this study. The question marks on Table 2 are listed for years where deaths are known to have happened, but that information is spotty and no firm number has been released, so this author decided to not list this incomplete data but also felt that the information was valuable for the reader to know that drone strikes did happen. It is also likely drone strikes happened in years where nothing is listed, and it is more than likely that drone deaths also happened in years where nothing is listed.

Global public opinion polls on US drone strikes

A July 2013 Pew Research Center survey found that the US military drone strike campaigns in countries like Pakistan, Yemen, and Somalia "are widely unpopular in the rest of the world" and that in 31 of 39 countries "at least half of the publics disapproved of the attacks" (Drake 2013; Pew Research Center 2013). While the Pew Research Center did not record data for every country from 2012–2014, one can see in Tables 3 and 4, that the numbers of disapproval grow and the percentage of approval declines. For example, in 2012 in the United Kingdom the disapproval rating was 47%, by 2013 it increased to 51%, and by 2014 it was up to 59% (Table 3).

Table 3. 2012-2014 Pew Research Center survey data with approval and disapproval of US military drone strike campaigns out of 100% (Pew Research Center 2019).

	2012	2013	2014
	Disapproved	Disapproved	Disapproved
Australia	-	48%	-
Canada	-	48%	-
New Zealand	-	-	-
United States	28%	30%	41%
United Kingdom	47%	51%	59%

Table 4. 2012-2014 Pew Research Center survey data with approval of US military drone strike campaigns out of 100% (Pew Research Center 2019).

	2012	2013	2014
	Approve	Approve	Approve
Australia	-	44%	-
Canada	-	43%	-
New Zealand	-	-	-
United States	62%	61%	52%
United Kingdom	44%	39%	33%

During the same time approval numbers were found to be very high for US military drone strikes with approval ratings in 2012 and 2013 in the 60% and 2014 in the 52% range (Table 4).

American public opinion polls on US drone strikes

The 2013 Pew Research Center found that the 61% approval rating of drone strikes varied slightly by the political party with 69% of Republicans approving, 59% of Democrats, and 60% of Independents (Pew Research Center 2013). By May 2015, the Pew Research Center had found that 58% of Americans approved of US drone strikes with 35% disapproving, and the survey also found that only 48% were "very concerned that U.S. drone strikes endanger the lives of innocent civilians, while another 32% said they were somewhat concerned about this (Pew Research Center 2015). It is easy to see why the numbers are high when no information is given out by the United Sates to its citizens. In 2016, President Obama stated, "publicly documenting civilian drone deaths would help the U.S. learn from mistakes, even if some casualties are unavoidable," however this public documenting never happened, and then with the Trump Administration the drone strike secrecy continued. Hina Shamsi, Director of the American Civil Liberties Union's National Security Project, wrote in opposition of President Trump's "decision to keep civilian drone deaths secret" and that "President Trump has already weakened rules that sought to limit civilian deaths caused by this country's illegal and immoral lethal force program, in which it kills suspects in places where we are not at war. This order now shrouds those killings in even greater secrecy" (Watson 2019).

Liberal/conservative newspapers

Newspapers tend to lean towards a political philosophical view such as liberal, conservative, independent, etc. depending on their ownership. Depending on the leanings of the newspaper it could support or oppose its own government, foreign governments, and military actions such as drone strikes.

Table 5. Presidents and Prime Ministers during the time frame of 2008–2019. Liberal administrations are highlighted.

	United States	Australia	Canada	New Zealand	United Kingdom
2008	Bush	Rudd	Harper	Clark/Key	Brown
2009	Obama	Rudd	Harper	Key	Brown
2010	Obama	Rudd/Gillard	Harper	Key	Brown/Cameron
2011	Obama	Gillard	Harper	Key	Cameron
2012	Obama	Gillard	Harper	Key	Cameron
2013	Obama	Gillard/Rudd/Abbott	Harper	Key	Cameron
2014	Obama	Abbott	Harper	Key	Cameron
2015	Obama	Abbott/Trunbull	Trudeau	Key	Cameron
2016	Obama	Trunbull	Trudeau	Key/English	Cameron/May
2017	Trump	Trunbull	Trudeau	English/Ardern	May
2018	Trump	Turnbull/Morrison	Trudeau	Ardern	May
2019	Trump	Morrison	Trudeau	Ardern	Johnson

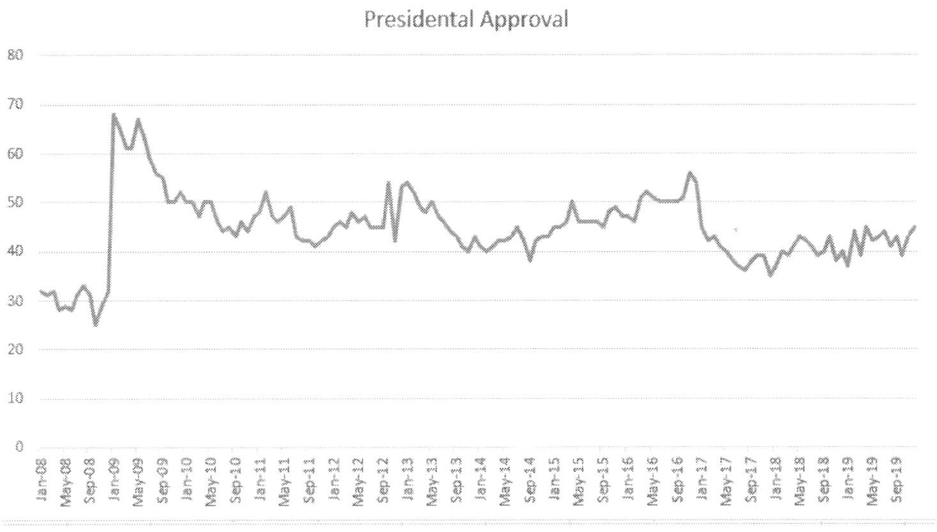

Figure 1. US Presidential Approval from The Gallup Organization for 2008-2019 reported by the Roper Center (Roper Center 2020).

Liberal/conservative governments (party in power)

Countries around the world have a variety of political parties, which have different philosophical or ideological views on how a government should be run. The political party in power will have an agenda for its relationship with other countries and global situations. Support for a country or military action might change as a political party takes power. The political parties during this study tend to fall into liberal and conservative ideology (Table 5).

In the United States during the timeframe of this study, 2008–2019, three different American presidencies held office: Republican George W. Bush (2001–2009), Democrat Barack Obama (2009-2017), and Republican Donald Trump (2017–2021) (Table 5 and Figure 1). They also had a wide variety of popularity or approval during their presidencies (Figure 1).

Structure of the Book

The book will consist of five main chapters. Chapter One - the Introduction, covers the state of the problem, the research question, the significance of the study, the key terms and definitions, the structure of the book, and the contributions to the discipline. Chapter Two - the Literature Review discusses drones, military drone strikes, and smart power theory. Chapter Three - the Methodology section, covers the research design, and the case study research, including the time period of the study, the data selected sources, data collection, and recording unit. Chapter Four covers the five case studies: The United States, Australia, Canada, New Zealand, and the United Kingdom and an analysis of the data. Chapter Five is the conclusion.

CHAPTER TWO

LITERATURE REVIEW

Drones

History of Drones

The history of drones can be seen to go as far back as balloons in the 18th century when Austrian hot air balloons bombed Venice, but hot air balloons are "generally not considered drones, mainly because their flight cannot be controlled" (Custers 2016, 9). Nikola Tesla, in 1897, created an early version of the "radio-guided electronic torpedo" and in 1898 "a radio-controlled motorboat" (Rinehart 2018, xi). In the United States in the 1910s, the Kettering Aerial Torpedo a.k.a. the Kettering Bug was constructed at the Dayton-Wright Airplane Company where inventor Charles F. Kettering, with the consultation of Orville Wright, helped create the first drone. The Kettering Bug was a bomb that would fall and explode on impact. The Queen Bee, the first radio-controlled biplane developed by the British between world wars, is often considered the first drone and is the origin of the term 'drone' (The Bureau of Investigative Journalism 2019). During WWII, both the United States and Germany experimented with remote-controlled airplanes, one even killing Joe Kennedy, the eldest brother of President John F. Kennedy in 1944 (Rinehart 2018, xi). With the rise of rockets and robotics following World War II, followed by the Space Race against the USSR and the Cold War, the United States began to "recognize the potential value of unmanned platforms as reconnaissance devices" (Springer 2013, 15).

Commercial Drones

Despite the usage of commercial drones by militaries and non-state actors around the world, this book will only look at the role of military drones and military drone strikes and those defined as military drones. However commercial drones are used by state militaries and non-state actors around the world due to their small size, low costs, availability, and relative ease of use. They are and will continue to be an important component in the coming years domestically and internationally and would be a wonderful topic for further study.

Military Drones

By the 1950s, Ryan Aeronautical Company (later part of Teledyne and later bought out by Northrop Grumman) began to build surveillance drones for the United States Air

Force. By the 1960s, the United States Air Force began using "an aerial target drone to train interceptor pilots and air defense personnel," called the Teledyne Ryan Firebee, which "followed a preprogrammed flight paths while carrying a camera" (Springer 2013, 15). Later, this early drone was renamed Ryan Model 147 Lightening Bug (later called Firefly), and it flew more than 3,400 reconnaissance missions over Vietnam (Springer 2013, 16). As the Vietnam War continued, other drone models were added for reconnaissance and surveillance over Vietnam, as well as other areas in Southeast Asia, China, and the Soviet Union (Springer 2013, 177–78). By the end of the Vietnam War, funding and development for drones began to dry up. However, by 1994, the United States Air Force and the CIA were looking at aerial reconnaissance and surveillance drones and agreed to "test the usefulness" of the General Atomics MQ-1/RQ-1 Predator as an "intelligence, surveillance, and reconnaissance (ISR) platform" (Springer 2013, 22).

Figure 2. MQ-1/RQ-1 Predator UAV. 2008a. United States Air Force photograph. Wikimedia Commons.

In 1995, the MQ-1/RQ-1[2] Predator (Figure 2) was being used over Yugoslavia and the Balkans for intelligence, surveillance, and reconnaissance (Springer 2013, 233). With a cruising speed of 90 miles per hour and costs around $5 million a drone, it had a range of 2,300 miles and can operate at 25,000 feet (Springer 2013, 191; Clausen 2017; Rinehart 2018, xxii).

The Predator's greatest value, Springer argues, is "the wide variety of sensor packages it can wield. They range from simple video cameras, including the one used by the pilot to

[2] The R stands for reconnaissance aircraft, the Q stands for unmanned vehicle, and the M stands for air-to-ground missiles (Springer 2013, 22).

fly the aircraft, to infrared cameras, synthetic aperture radars, and laser target designations" (2013, 191–92). Following 9/11, the Predator was armed with Hellfire missiles, and it is considered the first drone used for drone strikes, which were dropped in Yemen on November 3, 2002 (New America 2019; Clausen 2017; Rinehart 2018, ix). It was used in combat for over 21 years until its retirement in 2017 (New America 2019; Clausen 2017; Rinehart 2018, ix). The Predator was a very popular drone and became part of the government culture, so much so that at the 2010 White House Correspondence Dinner, President Obama joked during his speech that the "Jonas Brothers are here, they're out there somewhere. Sasha and Malia are huge fans, but boys, don't get any ideas. I have two words for you. Predator drones. You will never see it coming. You think I'm joking?" (Obama 2010). Under President Obama, drone use expanded for "surveillance and attack purpose in Yemen, Pakistan, and Afghanistan" (Whiton 2013, 51). Whiton writes, the CIA "became almost a second air force with its heavy usage of drones, especially to kill terrorists in Pakistan and Yemen" (2013, 164). The CIA, run its "bombing efforts in parallel to that of the military – namely, the Joint Special Operations Command" (Whiton 2013, 164).

Figure 3. MQ-9/RQ-9 Reaper. 2008b. United States Air Force photograph. Wikimedia Commons.

The General Atomics MQ-9/RQ-9 Reaper (Figure 3) "can be remotely controlled or flown autonomously...and can cruise at three times the speed of the Predator" and "according to former Air Force Chief of Staff Gen. T. Michael Mosely, the Reaper helped to transition UAVs from primarily "intelligence, surveillance, and reconnaissance roles" to "to a true hunter-killer role" (Stratfor 2019). The Reaper has been in operation for over 10 years, has nearly a 4,000-pound payload that can carry bombs and missiles, a cruising speed of over 240 miles per hour, can fly for 35 hours without refueling, and can fly above 50,000 feet (New America 2019; Clausen 2017; Springer 2013, 199). The Reaper's price tag for collection, acquisition, and targeting is about $14 million per drone (Rinehart 2018, xxii).

Figure 4. RQ-4 Global Hawk. 2007. United States Air Force photograph. Wikimedia Commons.

The RQ-4 Global Hawk (Figure 4), manufactured by Northrop Grumman, is a large drone created for high altitude and long endurance (Rinehart 2018, xxii). It is "capable of surveying as much as 100,000 square kilometers (40,000 square miles) of the landscape a day," and goes as fast as 310 knots, has a range of 16,000 miles, can fly as high as 65,000 feet, and costs around $225 million per drone (Stratfor 2019; Rinehart 2018, xxii; Springer 2013, 181).

Besides the three popular big drones of the Predator, Reaper, and Global Hawk, there are many smaller drones in use by the US military and they include the Fire Scout, the ScanEagle, the Puma, Switchblade, Wasp III, and the Raven (Rinehart 2018, xxii–xxiii).

Pros and Cons of Using Military Drones

Pros of using military drones

After September 11, 2001, military drones, like the Predator, found early success due to "operating in a highly permissive environment against an enemy with virtually no air defense network, demonstrated that under certain circumstances, unmanned platforms have potentially more utility than their manned counterparts" (Springer 2013, 23). Also, their "long loiter capabilities and slower flight speeds, meant a much greater capability to employ advance sensors against ground targets" (Springer 2013, 23). It's also helpful that drones like the Predator can operate at 25,000 feet (50,000 feet for the Reaper and 65,000 feet for the Global Hawk), and they are "virtually undetectable at night and fairly innocuous even in daylight" (Springer 2013, 23, 181).

Military drones are also advantageous due to 1) allowing for "sustained presence over potential targets," 2) providing "a near-instantaneous responsiveness – dramatically shrinking what the US military targeting experts call the 'find-fix-finish' loop – that most other platforms lack," and 3) most importantly "unmanned systems do not face the limitations associated with manned systems" such as the "risk of death or capture" (Kreps 2016, 25–26).

Cons of using military drones

The disadvantages to using military drones include that they "fly relatively slowly (the cruise speed of an F-16 is about three times that of a Reaper) and have therefore been described even by one Air Force general as "useless in a contested environment" (Kreps 2016, 38). Also, most drones, such as the Predator and Reaper are "sufficiently aerodynamic to meet the rigorous demands of air combat" (Kreps 2016, 151). Rinehart argues that there are many problems with using military drones: 1) "innocents are often killed, including American service people, and collateral damaged occurs," 2) a "lack of legal procedure that accompanies them. An individual is found guilty of committing an act of terrorism, yet evidence is never presented and there is never a trial," 3) "it is dehumanizing. Pilots are often flying these drones from the United States' military bases, their cockpits located in tractor trailers, without ever placing a foot in the war zone," 4) "numerous mechanical deficiencies in drones that cause them to crash, a lot," 5) "their astronomical cost," 6) they are a violation of state sovereignty," 7) "the concept of a 'legitimate target,' which is unclear and central to the debate of target killing, and 8) "blowback," which refers to the "unintended consequences of targeted killing" (2018, xiii-xxiii). There is also the issue of whether killing an individual by drone is an assassination or targeted killing. Grayson argues that there is a claim "that those states who label their assassinations or political executions as 'targeted killings' possess the technical and methodological acumen to obtain reliable intelligence, make decisions, and act upon them in a way that minimizes indiscriminate harms as outlined in humanitarian law." (2016, 4-5).

The price of being a pilot who is removed from the theater of war is taking its toll. A 2014 study of 1,000 United States Air Force drone operators found that 4.3 percent "of them experience moderate to severe PTSD" vs "10 to 18 percent of US military personnel returning from deployment" (Blaszczak-Boxe 2014). US military pilots are not the only ones dealing with the trauma of drone warfare, as the analysts watching and analyzing drone footage find themselves being "exposed to the most gruesome things that you can think about could happen on a battlefield...they find mass graves; they witness executions," as well as rapes, beheadings, and torture (McCammon 2017).

Military drone strikes have other social and cultural consequences in that they often lead to breeding "anger and frustration, leading to a backlash" against the US and other governments, such as in Pakistan (Jha 2014, 141). The Pakistan government was at least publicly displeased with US drones and drone strikes. Pakistan's Foreign Affairs Advisor to Prime Minister Sartaj Aziz told the *New York Times* in 2013 that "drone attacks are counterproductive to our relationship...we are asking for stopping" (Gordon 2013). They have led to a "number of suicide attacks in public places, including the one on Forward Operating Base Chapman in Afghanistan, in which seven CIA officers were killed" and "planting of a bomb in Times Square was also in retaliation to US drone killings" (Jha 2014, 142). Jha writes that retired Pashtun brigadier Mahmood Shah has publicly stated that "tribal code requires the families of drone victims to seek revenge, which makes them ideal recruits for insurgency and terrorist activities" (Jha 2014, 161).

Retired United States General Stanley McChrystal, who authorized the use of drone strikes in Afghanistan, has said that "what scares me about drone strikes is how they are perceived around the world...The resentment created by American use of unmanned strikes...is much greater than the average American appreciates. They are hated on a visceral level, seen by people who've never seen one or seen the effects of one" (Alexander 2013). He goes on to say that "the use of drones exacerbates a perception of American arrogance that says, 'well we can fly where we want, we can shoot where we want, because we can'" (Alexander 2013).

Military Drones and Drone Strikes

Drones are part of a technology that has been evolving at a rapid pace over the past few years. As Zenko and Kreps note, "the technology has evolved from surveillance aircraft to an armed platform, drones have been used for a wide range of military missions: The United States has successfully and legitimately used armed drones to conduct hundreds of counterterrorism operations in battlefield zones, including Afghanistan, Iraq, and Libya" (2014, 3). The United States has "also used armed drones in non-battlefield settings, specifically in Pakistan, Yemen, Somalia, and the Philippines" (Zenko and Kreps 2014, 3). The argument that will span the test of time is that drones and other robotics in "warfare will be driven by a desire to minimize the risk of one's own soldiers, but also a belief that such systems and their precision-strike capabilities will reduce the risk of 'collateral damage' among civilians" (Cavelty, Fischer, and Balzacq 2016, 460).

For many years, countries that used armed military drones were limited to the United States (first used in 2001 in Afghanistan), the United Kingdom (first used in 2008 in Afghanistan), and Israel (first used in 2004 in Gaza), but that changed in 2015 when Pakistan used a Burraq drone against individuals in Pakistan (New America 2019). Since then, Nigeria, Iran, Azerbaijan, Iraq, Turkey, and the UAE (against forces in Yemen and Libya) have joined the military drone strike community (New America 2019).

There is a growing group of countries that now have armed drones and mostly these drones are being developed and sold from China (New America 2019). Countries that are known to have armed drones are the United States, Israel, the United Kingdom, Iran, UAE, North Korea, China, South Africa, Italy, Nigeria, Spain, Pakistan, Iraq, Egypt, Georgia, Kazakhstan, India, Greece, Sweden, Switzerland, Ukraine, Turkey, Azerbaijan, France, Turkmenistan, Saudi Arabia, Poland, Taiwan, Belarus, Germany, Russia, Algeria, and the Netherlands (who are ordering them from the United States State Department in 2019) (New America 2019). The US military and CIA have used armed drones in Afghanistan, Iraq, Yemen, Pakistan, Somalia, Libya, and Mali (Jha 2014, 141). The British military has used armed drones in Afghanistan, Iraq, and Libya (Jha 2014, 141).

In the world of military drones, there is also a great rise in those countries that manufacture armed drones and sell them to other countries. The big three manufactures and sellers of drones are the United States, China, and Israel, but other countries have started to manufacture armed drones, including Australia, Belarus, France, Georgia, Germany, Greece, India, Iran, Italy, Jordan, North Korea, Pakistan, Russia, Saudi Arabia, South Africa, South Korea, Spain, Sweden, Switzerland, Taiwan, Turkey, UAE, Ukraine, and the United Kingdom (New America 2019).

Once someone has a drone, everyone wants a drone and that includes non-state actors. Hezbollah was the first non-state actor to deploy military surveillance drones against Israel in 2004 (New America 2019). ISIS in 2017 established a formal drone unit of their military – the 'Unmanned Aircraft of the Mujahideen' (New America 2019). The Houthi Rebels, a rebel group in Yemen backed and drone-supplied by Iran, in 2017, had a maritime drown strike against a Saudi warship that left two Saudi sailors dead and three injured (New America 2019). Hamas, a Palestinian group in Gaza, has Iranian-made surveillance, armed, and guided-missile drones (New America 2019). Other non-state actors that are known to have used commercial drones against their enemy's in combat zones include "Libyan rebels, the Donetsk People's Republic, the Kurdish Peshmerga, Jabhat al-Nusra, Faylaq al-Sham, and Saraya al-Khorani, both Syrian rebel groups, the Revolutionary Armed Forces of Colombia (FARC), and Colombian and Mexican drug cartels such as the Cártel de Jalisco Nueva Generación (CJNG), a group active in Central Mexico" (New America 2019).

Research on Drone Strikes

Because the US government does not release drone strike data, it has fallen on non-governmental organizations to research and report drone strikes and the number of individuals killed. The main non-governmental organizations that collect drone strike data are "The Bureau of Investigative Journals (BIJ), the New American Foundation, and the Long War Journal" with other organizations like "Amnesty International, Human Rights Watch, and Stanford/NYU Law" adding to the research with on-site interviews (Kreps 2016, 19–20). Photojournalists from the Associated Press, Getty Images, Reuters, and the Agence-France Press have managed to take photographs of the drone strikes, but this is often risky as the Taliban and the Pakistani Inter-Services Intelligence and the Ministry of Interior monitor strike zones, control the areas, and have been known to limit access and even punish local and international journalists for carrying a camera (Parks 2018, 147-148).

Academic research into the use of drones or robotics in warfare and their ethical and legal issues is a new topic and one that only started in 2009 when Arkin (2009), Krishnan (2009), and Singer (2009) all published books on drones and warfare and security (Cavelty, Fischer, and Balzacq 2016, 459). Since 2009, drone articles and books that deal with drone security studies fall into two categories: 1) drones and legal studies published by scholars and think tanks and is of a "military and strategic nature" and 2) studies that focus on drone warfare and target killings "an actual and observable practice of warfare using in-the-loop weapons, and is in the tradition of critical Security Studies" (Cavelty, Fischer, and Balzacq 2016, 459). For this book, articles and monographs from both categories were read to broaden the scope of the readings and knowledge for the book and the author.

There has been some literature on the legal and ethical use of drones by the military. Rinehart studied and wrote about drones and targeted killing in the Middle East and Africa (Rinehart 2018). Jha has written on the ethical, legal, and strategic implications of using drones in warfare and writes about the use of "drones perpetuate the illusion that military force is an effective means of countering terrorism. We should know better from the fact that more than a decade of combat in Afghanistan, has not eliminated the threat of terrorist attacks and insurgent violence in the region" (Jha 2014, 159).

Drone strikes and media coverage is also a new area of study by scholars. Cohen (2018) reviewed the origins of the United States' public support for drone strikes from 2000–2015 (Figure 5). White (2015) reviewed alternative (Amnesty International,

American Civil Liberties Union, CODEPINK, and Human Rights Watch) press publications, mainstream media coverage (*USA Today* and the *New York Times*), and drone strikes from 2009 to 2013 (Figure 5). In his review of alternative and mainstream media coverage, White found that "alternative media organizations primarily acted as voices of critiques from outside the standard of government silence that dominated the coverage of drone strikes from both mainstream media organizations" (2015, 342). He also found that alternative media broadens "the discussions about the issue of drone strikes from only accepting the information that came directly from government sources" and that "a high level of security emerged from the Obama administration on the topic of drone strikes" (White 2015, 353).

Sheets, Rowling, and Jones (2015) examined American, British, and Arab news coverage of US drone strikes for the years 2009–2012 (Figure 5) and found that "US news coverage was most favorable towards the drone program, emphasizing the strategic value, legality and technological capability of drones and downplaying civilian casualties," whereas Arab and British news outlets "were significantly more critical in their coverage of drones: Arab coverage was the most critical, sharply disputing drones' legality and technological sophistication. British news outlets fell somewhere in between the poles, with more mixed-valence coverage" (2015, 304). Interestingly, Sheets, Rowling, and Jones also found that "American news coverage was much less likely to emphasize civilian casualties, and that the overall valence of such coverage was much more pro-drone than coverage in the other outlets" (2015, 305). They believe and attribute this to the US news viewing public as being more favorable and uncritical of the US drone policy during the first Obama administration and that "Americans were not – and may indeed still not be – receiving an accurate picture of the ramifications and implications of the US drone policy" (Sheets, Rowling, and Jones 2015, 305).

Researchers	Title	Media type utilized	Newspaper Name	Years Covered	Documents Analyzed
Melissa K. Aho (2021)	"When You Play the Game Of Drones, You Win Or You Die: Examining The Role Of U.S. Drone Strikes In U.S. And English Language Allies Newspapers From 2008-2019"	10 National and International newspapers	The New York Times, USA The Wall Street Journal, USA The Age, AUS The Australian, AU The Globe and Mail, CA National Post, CA The Press, NZ New Zealand Herald, NZ The Guardian, UK The Times, UK	2008-2019	4,683 newspaper articles
Grant M Cohen (2018)	"Origins of the U.S. Public Opinion for Drone Strikes: The Intersection of Elite Rhetoric, Media Coverage, and American Public Opinion, 2000-2015"	News media and transcripts (media coverage), *Congressional Record* documents (elite rhetoric)	NBC, ABC, CBS, CNN, and FOX	2000-2015	576 total news recordings, 274 transcripts of news records, 452 *Congressional Record* documents
Brion White (2015)	"Killing in Silence: Alternative and Mainstream Media Coverage of Drone Strikes"	Newspapers: National and International alternative and mainstream media coverage	USA Today, New York Times, Amnesty International, American Civil Liberties Union, CODEPINK, and Human Rights Watch	2009 - 2013	88 from USA Today, 435 New York Times, 164 examples of media content from each of the four alternative media organizations
Penelope Sheets, Charles M Rowling, and Timothy M Jones (2015)	"The View from above (and below): A Comparison of American, British, and Arab News Coverage of US Drones."	News media: US, British, and Arab news coverage	CNN (US) BBC (British) Al Jazeera and Al Arabiya (Arab)	2009-2012	2,358 news stories

Figure 5. Comparison of drone strike research.

Newspapers and Media Bias

Newspaper and the media bias are as old as newspapers themselves. They often form a specific opinion on the matter, because despite how unbiased they claim, they do have biases that affect what is investigated, how it is edited, and what is published. Most newspapers tend to fall into two political affiliation groups of liberal and conservative (Worldpress.Org 2019). Bennett has written about such biases in the media and writes that "many politically heated stories raised troubling questions about what journalists should do when officials say things that are inconsistent with available evidence" (2016, 15). He believes that the American press has become dependent on "how those in power define reality" (Bennett 2016, 15). Bennett has even created a term for this brand of reporting – *indexing* (2016, 15–16). Bennett goes on to define *indexing* as "the tendency of mainstream news organizations to index or adjust the range of viewpoints in a story to the dominant positions of those whom journalists perceive to have enough power to affect the outcome of a situation" (2016, 16).

Newspapers are also about information and "politicians, businesses, and interest organizations that largely define politics in America, it has long been clear that power and influence depend on the control and strategic use of information" (Bennett 2016, 11). Diplomatic historian Marc Trachtenberg once wrote that it is important to remember that "documents are not necessarily to be taken at face value, and one has to see things in context to understand what they mean. One has to get into the habit of asking why a particular document was written – that is, what purpose it was meant to serve" (George and Bennett 2005, 107). Deborah Larson also argues that "it is important to understand the purpose of a document and the events leading up to it in order to correctly interpret its meaning" (George and Bennett 2005, 108). News accounts, argues Larson, "can help to establish the atmosphere of the times, the purpose of speeches or statements, or the public reaction to a statement. Newspapers help to show that information policymakers had and provide clues as to what events they regarded as important...In this way, newspapers help us to recapture the perspective of officials of the time" (George and Bennett 2005, 108). Being aware that bias exists is key to understanding newspaper articles and data and that a hidden agenda might be in play when using them in any study.

Smart Power Theory

This study will use the theory of smart power developed by Joseph Nye in 2003 "to counter the misperception that soft power alone can provide effective foreign policy" (Nye 2009a, 160) or by former Deputy Assistant Secretary of State Suzanne Nossel who used the term "'smart use of power' to promote U.S. interests through a stable grid of allies, institutions, and norms" (Wong-Diaz 2013, 22). Ambassador Paula Dobriansky writes that smart power is "synthesizing of diverse tools of statecraft, ranging from "soft" ones like foreign aid to "hard" ones like a military force. Essentially, it describes what government do when they bring multiple forces to bear on challenges from abroad" (Whiton 2013, ix). Power, Nye tells readers, "is one's ability to affect the behavior of others to get what one wants" and there are three basic forms "coercion, payment, and attraction" (Nye 2009a, 160). Smart power is defined as the "combination of hard and soft power," but there is more to it than that (Nye 2009b, 7). Smart power can be described as a "strategy that combines the soft power of attraction with the hard power of coercion" (Nye 2008, 1353). Wilson defines smart power as "the capacity of an actor to combine elements of hard power and soft power in ways that are mutually reinforcing such that the actor's purposes are advanced effectively and efficiently" (Wilson 2008, 110). Whiton writes that "at its most basic level, smart power should be thought of as progressing toward the foreign *political* outcome vital for American security and prosperity" (2013, xxi). In 2006, Nye and Richard Armitage co-chaired a commission to find a vision for "American's global engagement" (Wong-Diaz 2013, 23), and their final report joined "hard and soft power into a hybrid 'smart power' and recommended a smart power strategy that included: rebuilding and reinvigorating the foundation of alliances, partnerships, and institutions that serve our interests; elevating the role of development in U.S. foreign policy and developing a more unified approach to global development beginning with public health; public diplomacy directed at building people-to-people relationships, particularly with youth; economic integration and expansion of free trade; and addressing climate change and energy insecurity through technology and innovation" (Wong-Diaz 2013, 23).

Soft power "is the ability to obtain preferred outcomes through attraction rather than coercion or payments" (Nye 2009b, 7). Soft power includes a country's "culture (where it is attractive to others), its values (where they are attractive and not undercut by inconsistent practices) and policies (where they are not seen as inclusive and legitimate in the eyes of others)" (Nye 2009b, 7). Nye writes that an example of soft power is how "public opinion polls show a serious decline in American attractiveness in Europe, Latin America and, most dramatically, across the entire Muslim world" (Nye 2009b, 7). Soft power is also defined as "indirect or co-optive power behavior, the ability to get what you want by attracting and persuading others to adapt your goals" (Wong-Diaz 2013, 11). The "official instruments of soft power include public diplomacy, broadcasting, military-to-military contacts, exchange programs, development assistance, and disaster relief" (Wong-Diaz 2013, 11).

Hard power is the use of military force and coercion and is defined by Nye as "the use of coercion and payment" (2009a, 160). Hard power, writes Nye, is needed "to battle the extremists," while soft power is needed to "win the hearts and minds of the majority" (2008, 5). Finally, Wong-Diaz argues that "hard power is associated with realist and neorealist thinking and their strategic reliance on military force, economic sanctions, and coercive diplomacy as an instrument of statecraft," he goes on to write that "hard power is command power based on economic and military tools, and it is measured by such tangibles as population, military assets, GDP, natural resources, political stability, and so on" (2013, 11).

While the term smart power is new, the idea goes back generations. Back in the 1940s, the United States used a "multipronged smart power campaign against the rising tide of global communism" intellectually, economically, overtly, and covertly (Whiton 2013, xviii). A nation "can set the agenda for others or shape their preferences, it can save a lot on carrots and sticks. But rarely can it totally replace either. Thus, the need for smart strategies that combine the tools of both hard and soft power" (Nye 2009a, 160). Wilson argues that nations use "everything from smart bombs to smart phones to smart blogs" and that non-state actors, like Al Qaeda, use smart "media across multiple platforms" (Wilson 2008, 113). Any actor, writes Wilson, "that aspires to enhance its position on the world stage has to build strategies around the new fundamentals of 'smartness'" (2008, 113). President Obama in his 2020 autobiography discusses hard and soft power in his administration, writing that "even the more liberal members of his team...had no qualms about the use of "hard power" to go after terrorists" and "the most hawkish members of my team understood the importance of public diplomacy and considered the exercise of so-called soft power, like foreign aid and student exchange programs, to be essential ingredients in an effective U.S. foreign policy" (2020, 310–311).

Soft power is easily seen in the influence of the United States around the world especially in the years following the end of the Cold War, the American public's approval of anything military in the years following 9/11, the US military's continual explanation that drone strikes are precise and that they minimized the risk to soldiers and reduced collateral damage around civilians. Soft power can also be seen in the decline and opposition to the US military and its drones by citizens around the world and its English-speaking allies. Hard power is easily seen in the might of the US military as well as in the drones themselves with their increased firepower. Smart power then helps explain the use of military drone strikes around the world by the United States and why the United States English-speaking allies governments continue to support drone strikes, while the global public and their citizens do not. Drones and drone strikes need smart power to work and be accepted or acceptable.

CHAPTER THREE

METHODOLOGY

Methodology

This chapter discusses the methodology that was used in this study and how a content analysis approach and, more specifically, a case study approach was used to explore the role of newspaper media in the United States and English-speaking allies regarding their support or opposition to the US military drone strikes.

Content analysis can be defined as "the systematic, objective, quantitative analysis of message characteristics" (Neuendorf 2017, 1). It uses "messages rather than human beings as its units of data collection and analysis" (Neuendorf 2017, 36). Its job "summarizes rather than reports all the details concerning a message set" (Neuendorf 2017, 23). In content analysis, Berg tells us that seven major elements can be counted "words or terms, themes, characters, paragraphs, items, concepts, and semantics" (2009, 348).

The five case studies included in this study are the United States, Australia, Canada, New Zealand, and the United Kingdom. Open-source newspaper articles covering US military drone strikes from 2008 to 2019, covering twelve years, from a total of ten national and international newspapers were analyzed. Open-source newspaper articles were collected using these subject terms: 'drone strike' and 'United States,' as these are two basic subject terms that every relevant article would have. Other terms, such as terrorism or armed forces were not used as they are not very descriptive, and they do not get to the specific question being asked. Also, they would not have helped find articles about US drone strikes. Other terms for drones, such as 'Unmanned Aerial Vehicle' (UAV) or 'Uncrewed Aerial Vehicle' (UAV), or 'Unmanned Aircraft System' (UAS) are not common enough terms used by the general global population. To the general public, all drones are drones regardless of if they are military or civilian, and regardless of size or another technical element. From the selected newspaper articles of this study, a random sampling (created by using a random sequence generator) of ten articles will be selected for each year from 2008–2019 (Random 2021) (Figure 6).

	United States of America - The Wall Street Journal									
				Articles						
Year	1	2	3	4	5	6	7	8	9	10
2008 (20)	5	1	15	13	9	8	12	11	14	10
2009 (45)	7	22	40	45	28	39	2	23	17	19
2010 (102)	102	17	29	28	82	97	21	19	88	66
2011 (123)	65	20	66	97	75	78	50	101	69	111
2012 (99)	79	62	17	13	77	35	53	46	39	8
2013 (123)	114	95	106	28	101	65	99	9	49	35
2014 (89)	83	6	57	23	19	67	27	13	56	26
2015 (103)	101	38	78	60	71	79	47	69	8	70
2016 (56)	42	48	56	20	49	1	54	23	27	22
2017 (37)	10	14	6	13	12	29	3	1	37	34
2018 (36)	9	11	27	18	3	8	5	24	10	2
2019 (76)	35	19	50	24	26	54	3	14	63	17

Figure 6. Example of random numbers in the Wall Street Journal for 2008–2019. ProQuest US Newsstream.

Each of these ten random sample articles was examined and categorized as negative, positive, neutral, or no reporting for a content analysis approach and totaling 1,100 articles out of a total of 4,683 newspaper articles, had the potential to be reviewed by the author (Table 6). When the *Voyant Tools* software was used and analyzed, all 4,683 articles (Table 6) were used for each newspaper and each year.

Table 6. The 4,683-total number of articles used with the term 'drone strike' and 'United States' per year from 2008-2019 in the New York Times, the Wall Street Journal, The Age, The Australian, the Globe and Mail, the National Post, The Press, the New Zealand Herald, The Guardian, and The Times.

	2008	2009	2010	2011	2012	2013	2014	2015	2016	2017	2018	2019	Total
New York Times	23	107	127	166	199	228	183	172	120	139	75	161	1700
Wall Street Journal	20	45	102	123	99	123	89	103	56	37	36	76	909
The Age	3	4	8	16	23	21	23	16	7	9	3	24	157
The Australian	1	7	20	39	46	34	30	16	19	24	19	46	301
The Globe and Mail	4	9	15	27	16	17	14	18	7	15	9	25	176
The National Post	9	21	27	46	40	34	23	22	6	17	14	32	291
The Press	1	0	2	1	1	6	14	4	3	18	10	25	85
The New Zealand Herald	2	7	12	9	15	42	0	27	14	10	16	16	170
The Guardian	1	5	13	14	18	21	17	33	19	11	30	54	236
The Times	3	12	21	42	57	73	61	97	58	80	69	85	658
													4683

The open-source newspaper articles from the five countries were identified as possessing positive or negative themes towards US drone strikes in the articles. What defines the parameters for coding the articles as positive, negative, neutral, or no reporting is established by the author and shown below. Articles were briefly reviewed for themes and then entered in a spreadsheet after they were coded as positive, negative, neutral, or no reporting. Each of the ten newspapers had its own spreadsheet (Figure 6).

Articles that are coded as positive towards 'drone strikes' and 'The United States':
Articles that are coded as positive towards 'drone strikes' and 'The United States:' have these positive leanings/themes, that include the following:

- Emphasizing that drone strikes are consistent with international law
- Authorized by targeted country
- Effective tool for US military
- Save American lives
- Emphasis on strike accuracy
- Technological merits of drones in comparison to traditional warfare

- Minimize the number of civilian deaths caused by drone strikes or only mention militant causalities

Articles that are coded as negative towards 'drone strikes' and 'The United States':
Articles that are coded as negative towards 'drone strikes' and 'The United States' have these negative features/themes, that include the following:

- Strikes violate international law
- Not authorized by the country
- Futility of drones in fighting terrorism
- Cause Anti-American sentiment
- Inaccuracy of drones, technological merits overstated, or acknowledgment of civilian deaths caused by drone strikes

Articles that are coded as neutral towards 'drone strikes' and 'The United States':
Articles that are coded as neutral have no positive or negative leanings towards 'drone strikes' and 'The United States' or were articles that had both.

Articles that are coded as no reporting:
If the article cannot be used, such as not on topic (for example if the terms 'United States' or 'drone strikes' come up as a different topic than the one in this study) or if there was a duplicate article, they were coded as no reporting on the spreadsheet.

Table 7. Sample of the coding spreadsheet for newspapers in this study

United States of America - Wall Street Journal										
	Articles									
Year	1	2	3	4	5	6	7	8	9	10
2008										
2009										
2010										
2011										
2012										
2013										
2014										
2015										
2016										
2017										
2018										
2019										

Data mining with Voyant Tools

In the second level of analysis, the newspaper articles were uploaded to the text data-mining tool *Voyant Tools*. Data mining can be defined as "the process of analyzing large information repositories and through this method discover implicit but potentially useful information" (Gao 2017, 37). Data mining (also called text data mining, text mining, or text analysis) "is a way to analyze text data by identifying words and phrases to determine patterns or themes of the data set" (Gao 2017, 37). Text mining allows other patterns to be seen in newspaper articles that ordinarily would not be discernable to the human eye. The tools and software for data mining come in many different sizes and shapes, prices, and in the past have had to deal with complicated statistical and programming programs and languages such as Java and Python. In recent years with the rise of open-source and web-based tools and applications, data mining for text analysis has come into its own with user-friendly web-based, and open-source tools such as *Voyant Tools*, TagCrowd, VOSviewer, Overview, and Many Eyes.

Voyant Tools is an open-source project, user-friendly, and a scholarly project that combines a web-based text reading software and an analysis environment. It is "designed to facilitate reading and interpretive practices for digital humanities students and scholars as well as for the general public" (Sinclair and Rockwell 2016). Digital humanities scholars Stéfan Sinclair and Geoffrey Rockwell launched the software in 2003 to create a program for textual analysis that allows one to "see through your text" (Sinclair and Rockwell 2016; Sampsel 2018, 153). Daines, Nimer, and Lee agree that the "intent of textual analysis is to identify patterns and trends in a textual corpus" (2018, 5).

Data mining is not a new form of content analysis, but the use of *Voyant Tools* is in academics its early stages. Maramba et al. used *Voyant Tools* in their study of patient experiences surveys (2015), Marald et al. used *Voyant Tools* when they examined the "changing concepts in forestry journals in Sweden and the United States (US) in the early twentieth and early twenty-first centuries" (2016), and Cortes-Sanchez used *Voyant Tools* to examine the text of university mission statements to identify "characteristics related to language" (2018, 584). Campbell found that when looking at Tweets from the National Weather Service that *Voyant Tools* was scalable for large documents and found success with its word clouds, trends, and bubble line visual tools (2018, 31).

Voyant Tools, and other similar software that analyzes text, are at their most basic simply counting or sorting the frequency of words in a document; however, it does allow one to track word choices that rise and fall over time (Rockwell and Sinclair 2016, 34; Graham, Milligan, and Weingart 2016, 84). They can be sorted by frequency and "frequency counts can provide a sense of what a text is about" (Rockwell and Sinclair 2016, 34). Word clouds, another analysis tool and called 'Cirrus' in *Voyant Tools*, "provide a visually stimulating way of looking at a list of high-frequency words. The cloud arranges and sizes words by their rate of appearance. High-frequency words are made large and central in the cloud" (Rockwell and Sinclair 2016, 35). Graham, Milligan, and Weingart consider word clouds to be the simplest of the data visualization tools, and they let "you see a graphic representation of how often phrases appear in a body of text" (2016, 74-75). They argue that for historians, "the important element comes in change over time" and believe that "while they need to be used with caution, we believe that word clouds *can* tell us something" (Graham, Milligan, and Weingart 2016, 76-78). Another analysis tool is a distribution graph which displays the "locations of occurrences (or hits) in a text or corpus" which can "illustrate the rise and fall of a theme" (Rockwell and Sinclair 2016, 37).

Voyant Tools has many different tools to help the researcher analyze their text or corpus. Cirrus, Document Terms, Distinctive words, and Trends are used in this study. A Cirrus is a form of a word cloud that lets viewers visualize "the top frequency words of a corpus or document" (Voyant Tools 2020). Word clouds like Cirrus, are used to quickly draw the viewers' attention to those "high-frequency words" and are "best used in conjunction with other more exploratory and nuanced tools" (Voyant Tools 2020). The Distinctive words area "indicates the top five most distinctive words in each document" (Voyant Tools 2020). The Corpus Terms is "a table view of term frequencies in the entire corpus" (Voyant Tools 2020). Trends are line graphs that depict the "distribution of a word's occurrence across a corpus or document" (Voyant Tools 2020). The Trend tool shows a relative (term frequency) or a raw frequency that is the "absolute count for each document or document segment" (Voyant Tools 2020).

For this study, all the articles were saved in a PDF for each year, and then all the years for the specific newspaper were uploaded into *Voyant Tools*. For the twelve years studied, there are twelve documents. Duplicate articles, when found, were weeded out of the lists, as they were the same information and did not contribute anything new. A total of 4,683 newspaper articles were used for this part of the study (see Table 6).

The goal of text mining in these newspapers is to see 1) what terms occur throughout the newspapers, 2) is it possible to see themes appear or disappear over the time period of 2008–2019, and 3) are there any unique surprises found in the data.

Case Study Research

Yin writes that "data for case studies can come from many sources of evidence," and that there are six important ones "documentation, archival records, interviews, direct observation, participant-observation, and physical artifacts" (2002, 83). Documentation, which is used in this book, is "likely to be relevant to every case study topic. The type of information can take many forms and should be the object of explicit data collection plans" (Yin 2002, 85). Yin goes on to argue that "because of their overall value, documents play an explicit role in any data collection in doing case studies" (2002, 87). The documentation covers items like "newspaper clippings and other articles appearing in the mass media or in community newsletters," but it can also include items like administrating documents, formal studies or evaluations, letters, memoranda and other forms of communication, agendas, meetings minutes, and other reports or announcements that were written about events (Yin 2002, 85–86). Bruce Berg also defines case studies as "a method involving systematically gathering enough information about a particular person, social setting, event, or group to permit the research to effectively understand how the subject operates and functions," and with case studies a researcher can "capture various nuances, patters, and more latent elements that other research approaches might overlook" (2009, 317–18). The importance of newspapers cannot be overstated for the researcher. The National Library of Australia writes that newspapers are "primary sources, newspapers are of value to historians, researchers, academics and local and family historians, in that they not only record past events, but also provide a portal into different periods, cultures and societies throughout Australia's history. No other source can provide a record of our nation's reaction to disasters, triumphs, wars and droughts, while at the same time documenting the everyday lives of Australians...The fact that newspapers are compiled by people who are not consciously or deliberately recording for posterity, further benefits their value as sources by providing readers with a "birds eye view" of life as it was during a particular era" (2014).

Time Frame

This research covers the time frame of January 1, 2008 to December 31, 2019. In other words, it covers twelve years and three US presidents: George W. Bush, Barack Obama, and Donald Trump. This is so that a good representative of both conservative and liberal policies regarding the usage of drone strikes can be observed. Past studies on drone strikes and the media have covered periods from 2000 to 2015 (Cohen 2018), 2009 to 2013 (White 2015), and 2009 to 2012 (Sheets, Rowling, and Jones 2015).

Data Source Selection

Newspapers

Newspapers have been around since the 1600s and are notable because they try to send off a mass message to their readers. Mass messaging is "the creation of messages that are intended for a relatively large, undifferentiated audience" (Neuendorf 2017, 30). But the messages to the masses cannot get through if no one is reading them. In the 1980s, the "worst advertising recession in decades hit newspapers and continued for more than three years" and slowly in the 1990s, it grew until the hit of the internet, where one journalist called it "the biggest and saddest loser in the information revolution" (*The Economist* 1994, 41). In recent years, the newspaper industry has "been hit hard by the migration of readers to electronic media" (Hagan 2005). In fact, between 2004 and 2018, nearly 1,800 newspapers around the United States closed (Hagey, Alpert, and Serkez 2019). The *Wall Street Journal* and the *New York Times*, along with the *Washington Post*, found that between 2012 and

2018 their "circulation dropped an average of 29%" (Hagey, Alpert, and Serkez 2019). Major newspapers like the *Minneapolis Star Tribune* and *Chicago Tribune* found a 41% decline in that same timeframe and smaller midsize papers like the *Dallas Morning News* fell 45% (Hagey, Alpert, and Serkez 2019). Local newspapers are the hardest hit and have "suffered sharper declines in circulation than national outlets and greater incursions into their online advertising businesses from tech giants such as Alphabet Inc.'s Google and Facebook Inc.," and newspapers are also having "a more difficult time converting readers into paying digital customers" (Hagey, Alpert, and Serkez 2019). A "shrinking of the local news landscape is leaving Americans with less information about what's happening close to them," which even Facebook "acknowledged as it struggled to expand its local news product but couldn't find enough local stories" (Hagey, Alpert, and Serkez 2019). Consolidations and corporate takeovers are now seen as an industry norm. For example, the *Reading Eagle* founded in 1868 by the Barbey family recently declared bankruptcy when their online subscription service did not work. Publisher Peter Barbey said that the newspaper industry is "like we were in a canoe paddling upstream and there was a waterfall behind us" (Hagey, Alpert, and Serkez 2019). Even well-known businessman Warren Buffett believed that newspapers would turn around and began buying up local ones in 2011, but he has recently said that they are "toast" and that "with the exception of the *Wall Street Journal*, the *New York Times* and the *Washington Post*, 'they are going to disappear'" (Hagey, Alpert, and Serkez 2019). While newspapers across the country struggle to stay afloat, many believed that "even if newspapers vanished, journalism would flourish in a lower-cost, digitally native form. But those hopes have been misplaced" (Hagey, Alpert, and Serkez 2019). Fewer newspapers mean fewer jobs in the newsrooms, and a 45% decline in newsroom jobs from 2008 to 2017 (Hagey, Alpert, and Serkez 2019). Nicco Mele of the Shorenstein Center on Media, Politics, and Public Policy at Harvard University believes that a "future without newspapers...is actually a crisis for democracy" (Hagey, Alpert, and Serkez 2019). Many would argue that when there are fewer small and medium-sized newspapers, the news becomes one-sided, unpopular stories are lost, investigations are never started, and the voice of a few elite (be they individuals or governments) becomes the only message (Hagey, Alpert, and Serkez 2019; Herman and Chomsky 2002). Researchers have found that there is a dominant myth in our culture that our mainstream media has a liberal bias, when in fact they found that the opposite was true, and that the news has become subordinated to corporate interests and far to the right of most of the American population (Herman and Chomsky 2002). They also found that the media serves and "propagandize on behalf of, the powerful societal interests that control and finance them" (Herman and Chomsky 2002, xi).

Ten open-source newspapers were chosen for this research project because they are major and well-known newspapers in their home countries, that they have a liberal or conservative viewpoint, that they are English speaking or writing, and that the articles were accessible full-text for the whole time period covered in this book.

The newspapers that were used in this research are:

United States Newspapers
The *New York Times* (Liberal) and *Wall Street Journal* (Conservative)
Australia Newspapers
The Age (Liberal) and *The Australian* (Conservative)
Canada Newspapers
The *Globe and Mail* (Liberal) and *National Post* (Conservative)

New Zealand Newspapers
The Press (Liberal) and the *New Zealand Herald* (Conservative)
United Kingdom Newspapers
The Guardian (Liberal) and *The Times* (Conservative)

Data Collection Site and Procedures

Data in the form of open-source newspaper articles for the ten selected newspapers were collected from the newspaper databases *ProQuest US Newsstream* which "enables users to search the most recent premium US news content, as well as archives which stretch back into the 1980s," *ProQuest Canadian Newsstream*, which "offers unparalleled access to the full text of over 190 Canadian newspapers from Canada's leading publishers," and *ProQuest International Newsstream* "which provides the most recent news content outside the US and Canada" and "provides information from more than 660 of the world's top newspapers" were accessed from Southern New Hampshire University Library (ProQuest Canadian Newsstream 2020; ProQuest International Newsstream 2020; ProQuest US Newsstream 2020).

Each newspaper was searched with a date range of January 1 to December 31 for that specific year. While *Voyant Tools* removes stop-words like 'the,' 'and,' 'it,' etc. manually, it does not clean up and remove metadata information like column, publisher, place of publication, ISSN, page number, HTTP, HTTPS, times, etc.; in other words, it does not clean up words that are not relevant to the study. A list of stop-words was created, run through the articles numerous times, and then used on all of the articles.

CHAPTER FOUR

CASE STUDY RESEARCH AND ANALYSIS & RESULTS

Summary of Findings

Chapter 4 is a summary of the findings and the results to help answer the problems and the analysis of each of the five case studies. The five cases are the United States, Australia, Canada, New Zealand, and the United Kingdom and are listed below. These five countries also belong to an intelligence alliance called the Five Eyes (FVEY), whose origins go back to the post-WWII period and are part of the UKUSA Agreement. Edward Snowden called this alliance "a supra-national intelligence organization that doesn't answer to the laws of its own countries" (Seipel 2014). Investigative journalist Jeremy Scahill argues that the Five Eyes knew about the US drone strike policy and are not passive players (Freeze 2015).

This study asked the following questions:

1) Has American newspaper coverage of US drone strikes remained positive from 2008–2019?
2) Has coverage opposing US drone strikes increased in English-speaking newspapers of US English-speaking allies from 2008–2019?
3) Do the political leanings of newspapers influence the narrative for support for or opposition to US military drone strikes from 2008–2019?

Case 1 – United States

The United States has four significant political parties, and many smaller ones, which include the Democratic Party, the Republican Party, the Green Party, and the Libertarian Party (CIA 2019). The Democratic Party and the Republican Party are two major political parties in the United States. The government of the United States during the timeframe of this study, 2008–2019, includes three very different American presidencies: Republican George W. Bush (2001–2009), Democrat Barack Obama (2009–2017), and Republican Donald Trump (2017–2021) (Table 5).

Use of United States military drones and drone strikes

The use of military drones by the United States goes back to before 9/11, but for surveillance and intelligence gathering. The United States has had armed military drones since 2001. On October 7, 2001, the United States became the first country to use an armed drone, a Predator drone, in a drone strike in Afghanistan (New America 2019).

Newspapers United States

Two of the most important and popular national newspapers in the United States are the *New York Times,* a liberal newspaper, and the *Wall Street Journal,* a conservative newspaper. In this study, 1,700 articles were found in the *New York Times* and 909 articles were found in the *Wall Street Journal* under the study terms 'drone strike' and 'United States' for the years 2008–2019 (Figure 7).

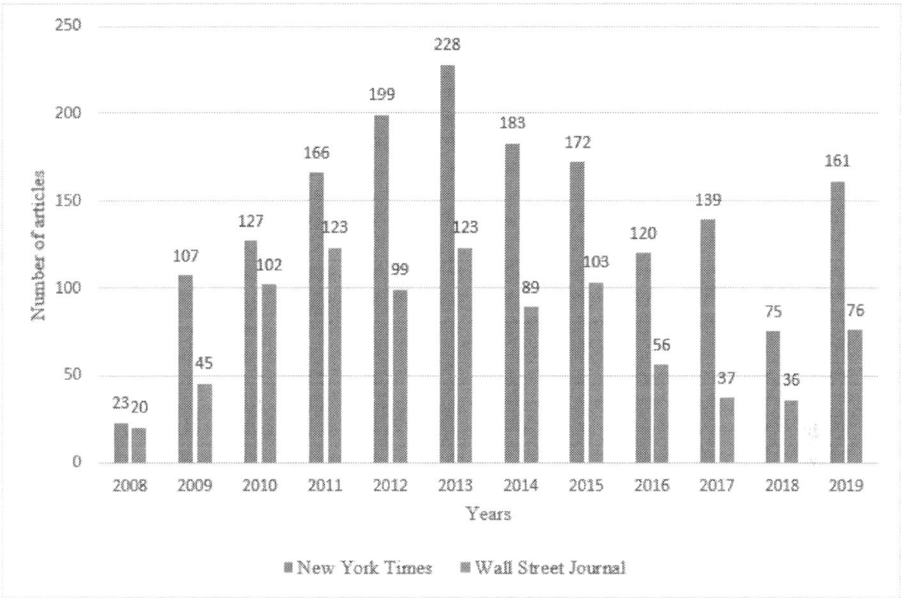

Figure 7. The number of total articles in the New York Times and Wall Street Journal with the terms 'drone strike' and 'United States' years of 2008-2019. ProQuest US Newsstream.

The New York Times (Liberal)

Figure 8. Cirrus graph of the twenty-five most commonly used words in the corpus of the New York Times from 2008–2019. ProQuest US Newsstream.

The *New York Times* based in New York City, New York started in 1851 as the *New York Daily Times* (*New York Times* 2019). It is a publicly traded company that is owned by the New York Times Company and the Sulzberger family. While no longer considered the newspaper of record, it has the second largest circulation in the 54, and has a liberal viewpoint.

In the 12 years for the *New York Times*, a total of 1,700 articles were found for a total of 2,688,528 total words, and out of those 51,412 unique words were found (Figure 9).

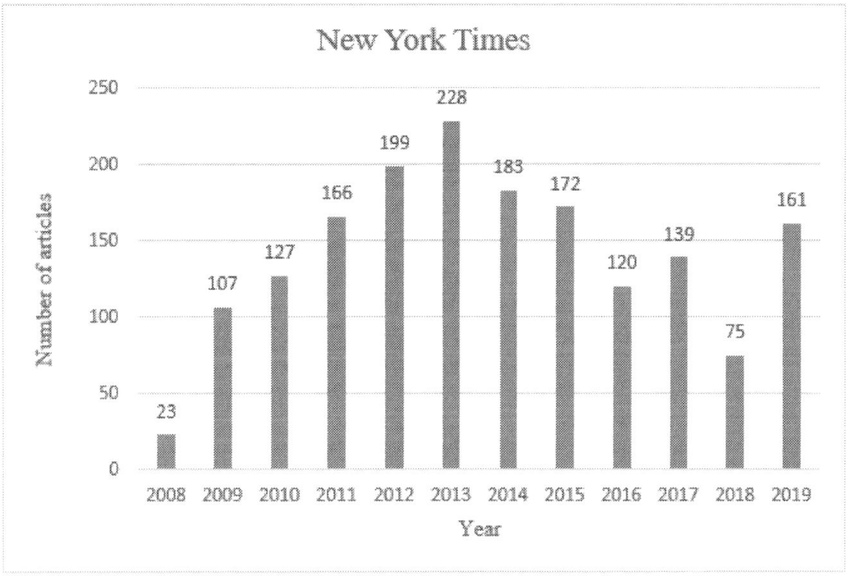

Figure 9. The number of total articles in the New York Times for the years of 2008–2019 with the terms "drone strike" and "United States." ProQuest US Newsstream.

The *New York Times* from 2008 until 2020 had more articles discussing US drone strikes than any other newspaper in this study. They peaked with 228 'drone strike' articles in 2013 and then descend, but even descending, the numbers remain high, such as in 2015 with 172 articles (Figure 9). The articles were slowly decreasing as the years went on, only to peak again during the Trump Administration in 2017. Interestingly, there is a bump in the number of articles in 2019 during the Trump Administration, but the numbers don't seem to rate a larger increase from previous years.

Positive articles were the majority of articles found in the sampling of 120 articles in The *New York Times* with 43 positive articles, 22 negative articles, 21 neutral articles, and 34 articles with no reporting (Figure 10). This is surprising as being a liberal newspaper, one would expect The *New York Times* to be more against drone strikes.

United States of America - The New York Times										
	Articles									
Year	1	2	3	4	5	6	7	8	9	10
2008 (23)	P	NE	NE	N	0	N	P	0	0	P
2009 (107)	P	P	NE	N	N	N	P	0	0	NE
2010 (127)	P	P	NE	NE	N	P	P	N	P	N
2011 (166)	N	0	P	P	P	P	P	0	P	P
2012 (199)	0	N	N	P	NE	0	0	P	NE	N
2013 (228)	0	N	NE	N	P	P	NE	0	N	N
2014 (183)	P	N	P	NE	N	P	NE	P	NE	NE
2015 (172)	P	NE	P	P	0	P	N	P	0	NE
2016 (120)	P	P	0	N	NE	NE	0	0	NE	N
2017 (139)	P	N	P	NE	P	P	0	P	0	P
2018 (75)	0	P	0	NE	P	0	P	P	0	0
2019 (161)	0	0	0	0	0	0	0	0	0	0

Figure 10. The New York Times Content Analysis for 2008–2019. Coding in figure: P = Positive, N = Negative, NE = Neutral, and 0 = No Reporting. ProQuest US Newsstream.

	Term	Count
1	united	11930
2	states	10956
3	american	10652
4	military	7068
5	officials	6362
6	pakistan	5527
7	obama	4692
8	drone	4407
9	afghanistan	4318
10	president	4156
11	state	4059
12	qaeda	4016
13	government	3910
14	intelligence	3878
15	people	3757
16	strikes	3661
17	taliban	3659
18	war	3611
19	security	3531
20	killed	3528
21	u.s	3502
22	coast	3501
23	pakistani	3406
24	country	3358
25	forces	3315

Figure 11. Top twenty-five most commonly used words in the corpus of the New York Times from 2008–2019. ProQuest US Newsstream.

The Cirrus graph of the 25 most commonly used words in the 1,700 articles from 2008–2019 find that words such as 'United,' 'States,' 'American,' 'military,' 'officials,' and 'Pakistan' are words that tend to rank the highest (Figure 8). The trend graph for the *New York Times* gives us a more detailed glimpse of the number of times a term has been used. President Obama is the only president listed, coming in a number 7 in the ranking. 'Pakistan,' ranked number 6, ranks higher than 'Afghanistan,' even though more drone strikes have occurred in Afghanistan. 'Drone' ranks eighth on the list, 'strikes' are listed at number 16 (the singular 'strike' does not make the top 25, 'war' is listed at number 18, and 'killed' at number 20. Negative terms such as 'war' and 'killed' do appear in the top 25 of the most commonly used words.

Distinctive words (compared to the rest of the corpus):

1. 2008: saakashvili (15), bajaur (33), rara (11), cosa (11), aurakzai (11).
2. 2009: atiqullah (31), tahir (38), tayyeb (25), zardari (78), munadi (16).
3. 2010: pence (268), quijano (112), kaine (232), shahzad (219), shahzad's (44).
4. 2011: libya (217), cuspert (24), kasinof (31), pacifists (18), qaddafi (144).
5. 2012: romney (165), polio (80), pimentel (35), qaid (23), declan (131).
6. 2013: album (79), kiriakou (45), tsarnaev (68), polio (45), j.p (25).
7. 2014: isis (362), pillay (26), kurdish (152), maliki (99), gaza (172).
8. 2015: isis (198), kouachi (55), marquez (37), emwazi (58), houthis (191).
9. 2016: rahami (105), artan (44), duckpin (37), isis (168), adnani (50).
10. 2017: abedi (78), isis (327), trump (446), trump's (118), yelp (41).
11. 2018: maven (39), nigerien (52), trump (198), isis (104), wakabayashi (20).
12. 2019: trump (1,268), pompeo (215), bolton (130), trump's (246), esper (72).

Figure 12. Most distinctive used words in the corpus of the New York Times from 2008–2019. ProQuest US Newsstream.

The most distinctive words used in the entire selection of 1,700 articles from 2008–2019 give us a glimpse into unique words occurring when compared against the whole corpus (Figure 12). Most of these terms do not make the top 25 or even the top 50 but appear due to a current event and then don't appear again. Countries like Libya appear in 2011 probably due to the assignation of its leader, terrorist organizations such as ISIS appear in 2014, and individuals appear such as the arrival of President Trump in the news 2017 and stay in the news in 2018 and 2019.

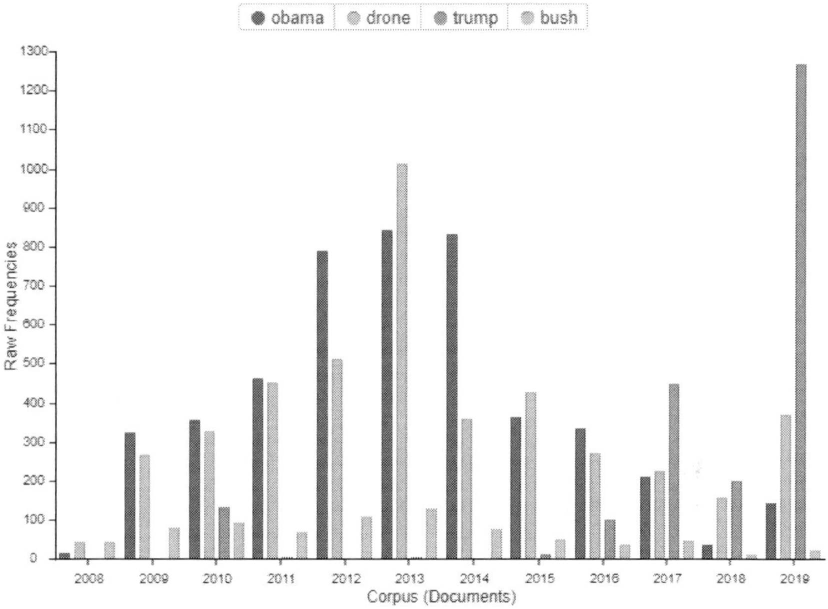

Figure 13. Trend graph on the raw frequency of the terms Bush, Obama, Trump, and drone used in the corpus of the New York Times from 2008–2019. ProQuest US Newsstream.

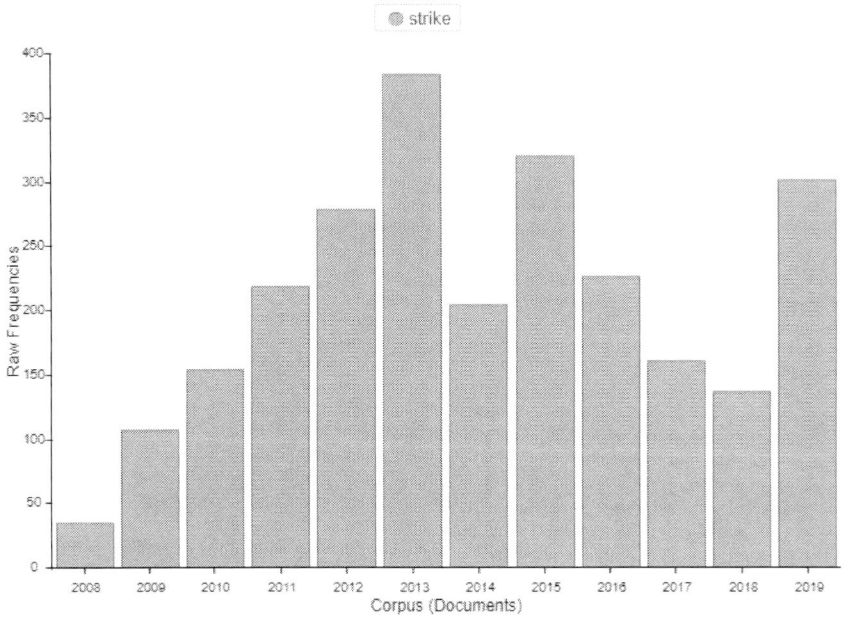

Figure 14. Column graph for the term strike used in the corpus of the New York Times from 2008–2019. ProQuest US Newsstream.

Presidents in the news with their relative frequency of the number of times their name has appeared show their arrival in the news and how they fade away from the story. President Obama is mentioned 4,692 times, President Trump is mentioned 2,159 times, and President Bush is mentioned 753 times, and the drone was mentioned 4,407 from 2008–2019 (Figure 13). President Trump was mentioned more times, in 2019, than any of the other presidents during any other year. The term 'strike' appears 2,523 times in the corpus of the *New York Times* from 2008–2019 and peaks in 2013 with the term appearing 383 times, and this was also the same year that it published the most articles with the term 'drone strike' (Figure 14). The year 2015 comes in at second place with 320 mentions, and 2019 comes in third place with 301 mentions. The term 'drone' appears most frequently in the year 2013, only to decline in numbers, until 2019.

The Wall Street Journal (Conservative)

Figure 15. Figure Cirrus graph of the twenty-five most commonly used words in the corpus of the Wall Street Journal from 2008–2019. ProQuest US Newsstream.

The *Wall Street Journal* based out of New York City, NY, boasts the third largest circulation in the United States and is published by Dow Jones & Company Inc. (Figure 15). Founded in 1889, it has become more conservative since it was purchased in 2007 by News Corporation, which is owned by conservative Australian-American businessman Rupert Murdoch.

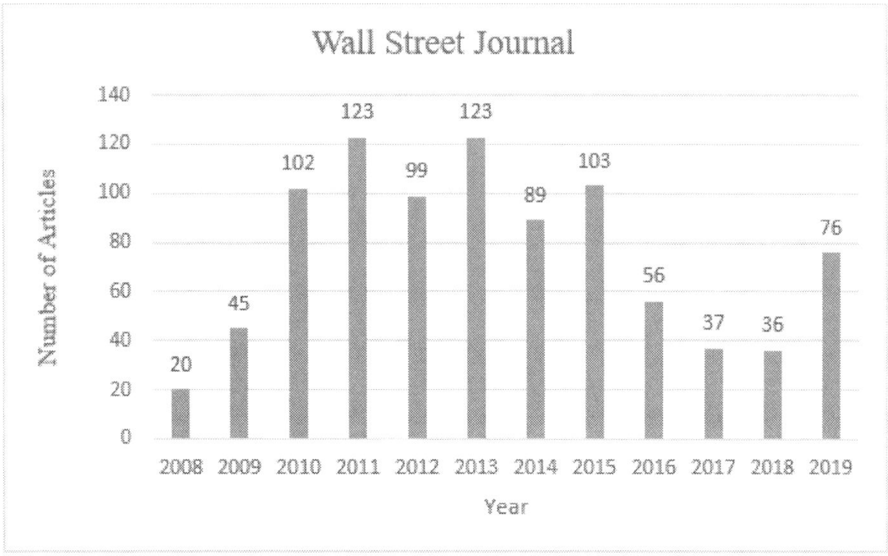

Figure 16. The number of total articles in the Wall Street Journal for the years of 2008–2019 with the terms 'drone strike' and 'United States.' ProQuest US Newsstream.

The *Wall Street Journal* had 909 articles with a total of 1,105,887 total words and 32,362 unique words with just the terms 'drone strike' and 'United States' (Figure 16). Overall, the *Wall Street Journal* has 53% fewer articles on drone strikes than the *New York Times*.

The peak number of drone strike articles in the *Wall Street Journal* is seen in 2011 and 2013 during the time of the US and NATO drawdown and the Taliban resurgence, only to peak again during the Trump Administration (Figure 16). This is different than the liberal newspaper the *New York Times*, which peaks in 2013, so it is peaking earlier. After 2015 with its decline in numbers, we see further declines in 2017 and 2018, only to see the number of articles increase and almost double what it was the year previously so that 2019 has 76 articles. So, there is a slight bump during the extreme drone strikes of the Trump administration, but it is not as high as in past years.

The reporting on the articles in the *Wall Street Journal* showed a lot of positive articles (Figure 17). There were 54 positive articles, 17 negative, 15 neutral, and 34 no reporting articles. Many of these positive articles had to do with Libya and the NATO and US drone activities. No reporting articles had to do with other topics, such as a Venezuela drone, Israeli and Iranian drones, Turkish drones, and Iran shooting down a US drone.

	United States of America - The Wall Street Journal									
	Articles									
Year	1	2	3	4	5	6	7	8	9	10
2008 (20)	N	0	N	N	P	P	N	P	0	P
2009 (45)	P	0	P	P	P	N	N	P	P	P
2010 (102)	N	NE	P	P	P	NE	P	P	P	P
2011 (123)	P	0	P	P	N	P	P	P	P	P
2012 (99)	N	P	0	0	N	NE	0	NE	P	0
2013 (123)	0	NE	N	0	NE	0	NE	N	0	N
2014 (89)	N	0	P	0	0	P	P	0	P	P
2015 (103)	NE	NE	N	P	NE	P	P	P	P	NE
2016 (56)	0	P	NE	P	P	0	NE	P	N	P
2017 (37)	0	0	P	P	P	P	P	P	P	N
2018 (36)	P	0	0	P	NE	0	0	NE	P	P
2019 (76)	0	0	0	0	0	0	0	0	0	0

Figure 17. The Wall Street Journal Content Analysis for 2008–2019. Coding in figure: P = Positive, N = Negative, NE = Neutral, and 0 = No Reporting. ProQuest US Newsstream.

	Term	Count
1	u.s	10336
2	wall	5622
3	officials	3638
4	military	3497
5	pakistan	2710
6	world	2504
7	united	2445
8	states	2222
9	drone	2113
10	journal	1975
11	obama	1924
12	intelligence	1854
13	state	1802
14	forces	1796
15	permission	1712
16	qaeda	1711
17	strikes	1670
18	afghanistan	1608
19	government	1584
20	country	1570
21	american	1517
22	taliban	1513
23	people	1501
24	president	1491
25	pakistani	1441

Figure 18. Top twenty-five most commonly used words in the corpus the Wall Street Journal from 2008–2019. ProQuest US Newsstream.

Where the *New York Times* had the most popular words in 'United,' 'States,' 'American,' 'military,' 'officials,' and 'Pakistan,' in the *Wall Street Journal* we see most of these terms further down from the top 25 list and words such as 'Obama' is further down on the list, but once again he is also the only president to make the top 25 (Figure 18). 'Drone' ranks on the list at 9, 'strikes' is listed at number 17 (the singular 'strike' does not make the top 25), and 'war' and 'killed' do not make the top 25. One unique term to make the top 25 is the term 'permission' which comes in at 15. Not using negative words like 'war,' 'killed' and using positive terms like 'permission' concerning the words 'drone strikes' gives authorization for their usage and shows that the *Wall Street Journal* is more supportive of this military action.

Distinctive words (compared to the rest of the corpus):

1. 2008: schuman (18), dreazen (31), yochi (27), zahid (34), 20dreazen (14).
2. 2009: mehsud (67), dreazen (42), 148 (149), ocampo (16), yochi (28).
3. 2010: shahzad (61), gorman (94), mehsud (53), rosenberg (49), sidiqi (17).
4. 2011: seychelles (31), entous (129), wright (58), anonymous (192), gigot (20).
5. 2012: romney (78), wright (50), benghazi (43), entous (80), knickmeyer (24).
6. 2013: aqim (42), irs (39), belmokhtar (37), subscribe (45), mali (107).
7. 2014: isis (155), dam (41), godane (27), maliki (73), mosul (44).
8. 2015: weinstein (86), tikrit (48), porto (27), houthis (88), christians (41).
9. 2016: raphel (35), donati (28), shenzhen (19), raphel's (18), jessica (30).
10. 2017: trump (73), kesling (19), gerasimov (18), affidavit (18), mosul (28).
11. 2018: ai (70), trump (79), kaplan (16), sune (13), somaliland (8).
12. 2019: trump (360), aramco (45), bolton (62), sune (43), pompeo (61).

Figure 19. Most distinctive used words in the corpus of the Wall Street Journal from 2008-2019. ProQuest US Newsstream.

The most distinctive words used in the entire selection of 909 articles from 2008–2019 highlight terror organization 'ISIS,' which appears in 2014; cities like 'Tikrit' and 'Benghazi,' and individuals like President Trump appear in the corpus in 2017, 2018, and 2019 (Figure 19).

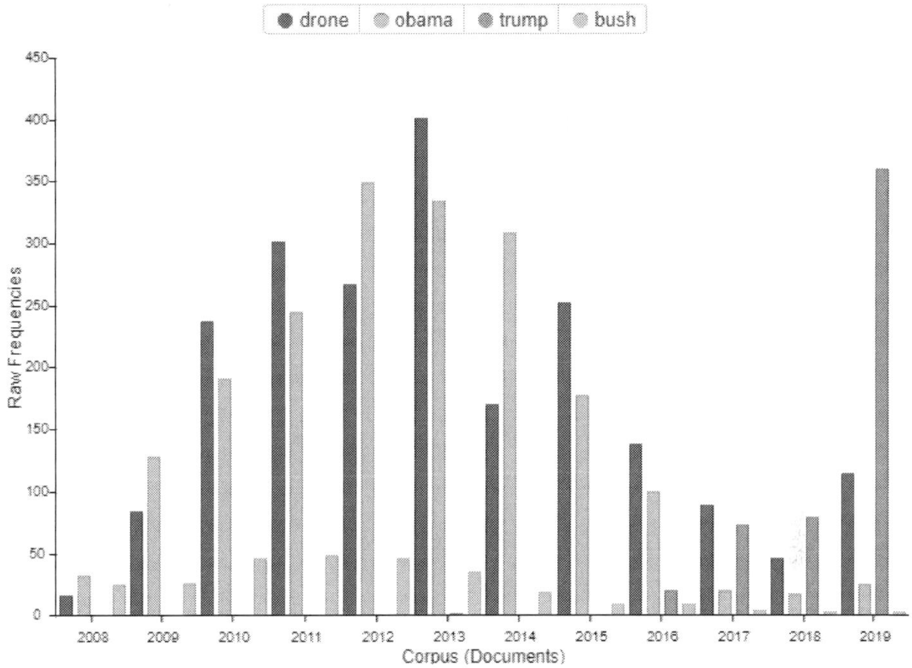

Figure 20. Trend graph on the raw frequency of the terms Bush, Obama, Trump, and drone used in the corpus of the Wall Street Journal for 2008–2019. ProQuest US Newsstream.

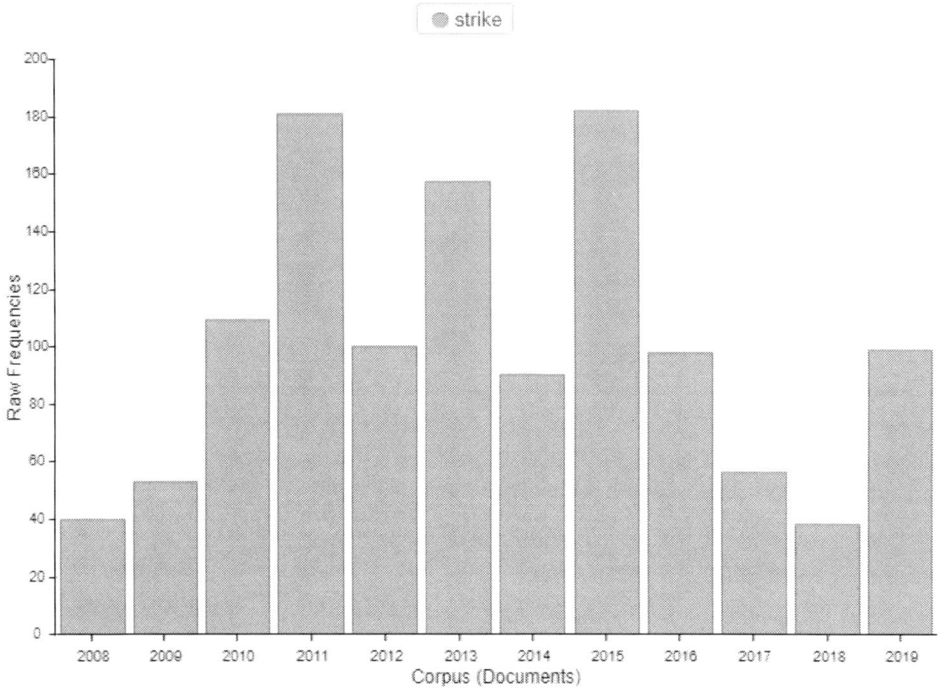

Figure 21. Column graph for the term strike used in the corpus of the Wall Street Journal from 2008–2019. ProQuest US Newsstream.

Much like the *New York Times*, in the *Wall Street Journal* the US presidents and their relative frequency concerning drone strikes and how quickly that frequency fades when they are no longer in the White House. President Obama is mentioned 1,924 times, President Trump is mentioned 533 times, and President Bush is mentioned 269 times from 2008–2019, whereas drone is mentioned 2,113 times during the same time period. President Trump was mentioned more times, in 2019, than any of the other presidents during any other year, although President Obama was mentioned three times more often overall in the 12 years (Figure 20). In 2010 and 2011 during President Obama's troop surge, the number of drone strike articles by both the *New York Times* and *Wall Street Journal* were not at their highest. The term 'strike' is mentioned 1,203 with the peak year being 2015 with 182 mentions, then 2011 with 181 mentions, and 2013 with 157 mentions, and by 2019 there are only 99 mentions (Figure 21). The frequency goes down after the peak in 2015 and only the new Trump administration, and the extraordinary number of drone strikes brought the numbers up again, but not to numbers seen in earlier years of this study.

Case 2 – Australia Newspapers

Australia has eight significant political parties, and many smaller ones, which include the Australian Greens Party, Australian Labor Party, Country Liberal Party, Liberal National Party of Queensland, Liberal Party of Australia, the Nationals, the Centre Alliance, and Pauline Hanson's One Nation (CIA 2019). Often in Australia, there will be a coalition where two parties will group together to win power. There is currently a coalition of the Liberal Party, and the National Party is in power. The coalition of the Liberal Party and the National Party is considered center-right, and the Labor Party is considered center-left. The Labor believes that "government must play a vital role in ensuring the public welfare, and it has strongly supported gender and radical equality and Aboriginal rights. The party has also favoured a more independent foreign policy and endorsed the idea of Australia becoming a republic" (Encyclopaedia Britannica 2019a). The 2007 elections were tough for the Liberal Party due to the "growing Australian concern over environmental issues, public services, and the country's involvement in the US-led Iraq War (Encyclopaedia Britannica 2019c). The political parties that governed Australia during the time frame covered in this research, 2008–2019 were Kevin Rudd (2007–2010), Julia Gillard (2010–2013), and Kevin Rudd (2013–2013) all from the Labor Party and then the Liberal coalition took over with Tony Abbott (2013–2015), Malcolm Turnbull (2015–2018), and Scott Morrison (2018–2022) (National Archives of Australia 2019) (Table 5).

Use of Australian military drones and military drone strikes

Australia has never used a drone in combat and never had any military drone strikes (New America 2019). However, Australia does seem to be changing its stance on military drones, and in 2017, a US Reaper drone from Boeing was shipped to Australia (New America 2019). The country has expressed interest in purchasing both US and Israeli drones (New America 2019). In 2019, the Australian Defense Force Academy, the University of New South Wales, and the University of Queensland were funded AUD $9 million for a project that looks at "how to make autonomous weapons such as future armed drones behave ethically in warfare" (Wroe 2019). In 2020, Boeing created a 'Loyal Wingman' drone for the Australian Air Force, for which Prime Minister Scott Morrison said that it would be "pivotal to exploring the critical capabilities our Air Force needs to protect our nation and its allies into the future" (Insinna 2020).

Newspapers in Australia

The earliest newspaper in Australia, the *Sydney Gazette and New South Wales Advertiser*, was established in 1803 and was "the only publication in the colony in a time of government censorship until William Wentworth launched *The Australian* in 1824" (National Library of Australia 2014). As with the rest of the world, newspapers in Australia see a decline in their circulation numbers. For the first six months in 2018, *The Age*, the *Sydney Morning Herald*, and the *Australian Financial Review* "fell by at least 10%" (Samios 2018). Other newspapers around Australia also saw declines regardless of their size (Samios 2018). In this study, a total of 157 articles for *The Age* and 301 articles for *The Australian* with the study terms of 'drone strike' and 'United States' for the years 2008–2019 were found (Figure 22).

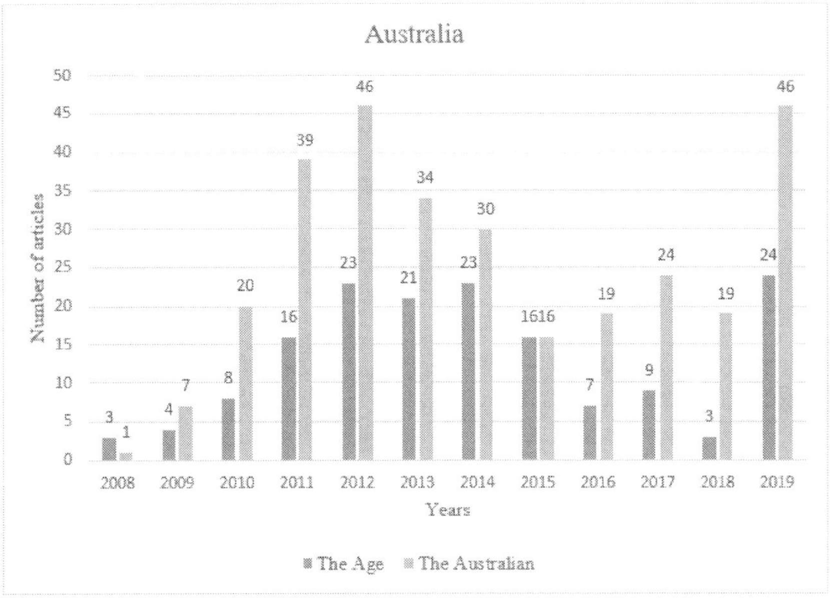

Figure 22. The total number of articles in The Age and The Australian with the terms 'drone strike' and 'United States' years of 2008–2019. ProQuest International Newsstream.

The Age (Liberal)

Figure 23. Cirrus graph of the twenty-five most commonly used words in the corpus of The Age from 2008–2019. ProQuest International Newsstream.

The Age, a liberal or center newspaper, is based out of Melbourne, Australia, and is owned by Fairfax Digital. From 2008–2019, *The Age* published a total of 157 articles with the terms 'drone strike' and the 'United States.' These articles had 193,607 total words and 15,472 unique words (Figure 23).

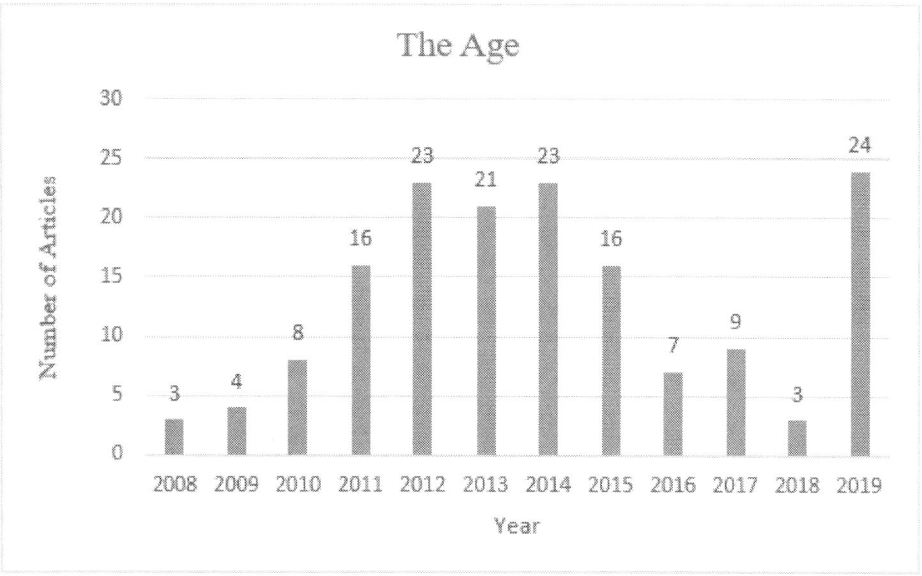

Figure 24. The number of total articles in The Age for the years of 2008–2019 with the terms 'drone strike' and 'United States.' ProQuest International Newsstream.

Of the 157 articles found for *The Age*, the article's year with the highest number of articles is 2019 with 24 articles (Figure 24). Otherwise, the graph shows that 2012–2014 have relatively the same numbers of 23, 21, and 23 articles respectively. After 2014, we see a decline in articles that continues until 2017, which then goes slightly up only to decline again in 2018 with 3 articles and then soars with 24 articles in 2019 during the Trump administration's new drone strike policy and the extremely high numbers associated with it.

The Age had only 6 positive articles, out of 94 total articles, with 31 negative, 21 neutral, and 36 with no reporting (Figure 25). No reporting articles include items like television programs with drones, drones by other countries like Israel and Russia, the selling of Predator drones to the UAE, drones that the Australian government and military want to use, or purchase, and some have nothing to do with US military drones. Negative articles dominate until 2013 and then articles fall under the no reporting categories. One example of a negative article from 2011, 10 years into the war, discusses the US and how "they have displayed an identical disregard for basic humanity: think Abu Ghraib, Guantanamo Bay and the widespread slaughter of innocents ("collateral damage") through the use of drone strikes" (Oborne 2011). Or as a 2012 article stated, "Drone warfare leaves the US President with the blood of innocents on his hands...If victims of Obama's drone strikes are mentioned by the state at all, they are discussed in terms that suggest they are less than human" (Monbiot 2012).

	Australia - The Age									
	Articles									
Year	1	2	3	4	5	6	7	8	9	10
2008 (3)	N	NE	0							
2009 (4)	N	N	P	N						
2010 (8)	N	0	N	N	N	N	P	N		
2011 (16)	NE	P	P	N	N	NE	NE	NE	P	N
2012 (23)	N	NE	N	NE	N	N	N	0	N	N
2013 (21)	N	N	NE	0	0	NE	N	N	N	N
2014 (23)	NE	N	NE	NE	N	NE	0	NE	0	NE
2015 (16)	NE	N	NE	0	0	0	0	0	0	0
2016 (7)	0	0	NE	0	0	0	P			
2017 (9)	0	0	NE	N	N	0	0	0	NE	
2018 (3)	NE	0	0							
2019 (24)	0	0	0	0	0	0	0	0	0	0

Figure 25. The Age Content Analysis for 2008-2019. Coding in figure: P = Positive, N = Negative, NE = Neutral, and 0 = No Reporting. ProQuest International Newsstream.

	Term	Count
1	military	426
2	war	414
3	international	332
4	pakistan	313
5	drone	311
6	obama	306
7	strikes	304
8	people	291
9	country	287
10	united	285
11	iraq	283
12	afghanistan	270
13	forces	265
14	air	264
15	states	249
16	general	247
17	defence	245
18	government	230
19	president	228
20	state	227
21	security	221
22	drones	218
23	qaeda	206
24	world	192
25	killed	191

Figure 26. Top twenty-five most commonly used words in the corpus of The Age from 2008–2019. ProQuest International Newsstream.

The terms that appear the most in articles from *The Age* are 'military,' 'war,' 'international,' 'Pakistan,' and 'drone.' 'Drone' makes the top 5 on the list ('drones' comes in at number 22), 'strikes' is listed at number 7 (the singular 'strike' does not make the top 25), 'killed' ranks in at 25 and 'war' is ranked number 2 (Figure 26). With the terms 'military,' 'war,' and 'international' in the top three, these terms are constantly used and with 'war' ranked in at number 2, they are given a negative connotation. 'Iraq' is even mentioned at number 11 with *The Age.*

Distinctive words (compared to the rest of the corpus):

1. 2008: gaza (25), ud (6), sderot (6), o'loughlin (6), jamaat (6).
2. 2009: anne (15), 20davies (8), mcchrystal (13), nicholson (6), davies (9).
3. 2010: fitzroy (39), jazz (36), rock (36), pop (43), hotel (46).
4. 2011: mann (27), libyan (19), gaddafi (31), bin (53), pimentel (11).
5. 2012: polio (37), doherty (29), romney (15), brody (15), kirner (10).
6. 2013: sharif (17), scahill (15), purcell (13), gibney (13), manning (26).
7. 2014: isil (27), baghdad (20), boko (18), abbott (22), haram (16).
8. 2015: singer (29), singer's (12), abbott (19), syria (60), sharrouf (9).
9. 2016: prakash (13), ball (17), twain (8), sdsc (7), rail (6).
10. 2017: trump (36), trump's (22), virtual (11), tillerson (8), sport (6).
11. 2018: pyne (5), privacy (5), miller (6), rhyolite (3), tritons (4).
12. 2019: trump (78), strait (19), needham (19), bachelard (26), iranian (49).

Figure 27. Most distinctive words used words in the corpus of The Age from 2008–2019. ProQuest International Newsstream.

The most distinctive words used in the 157 articles from 2008–2019 offer a view of those headline-making incidents that would appear in the news, only to leave once they were no longer current, as if they were popular, they would rank in the top 25 terms (Figure 27). The terrorist organization 'ISIL' appears in 2014 and is used instead of 'ISIS,' which is a more popular word choice in the American newspapers, and President Trump's name appears in 2017 and 2019 as distinctive words.

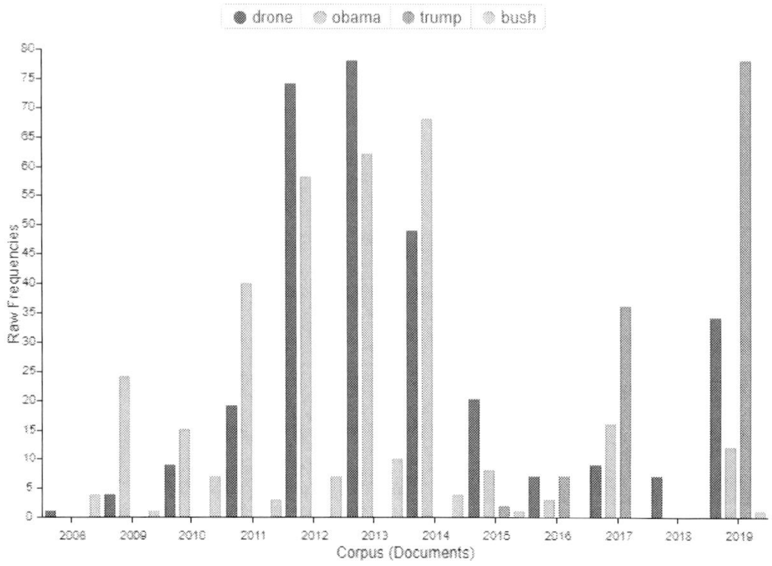

Figure 28. Trend graph on the raw frequency of the terms Bush, Obama, Trump, and drone used in the corpus of The Age from 2008–2019. ProQuest International Newsstream.

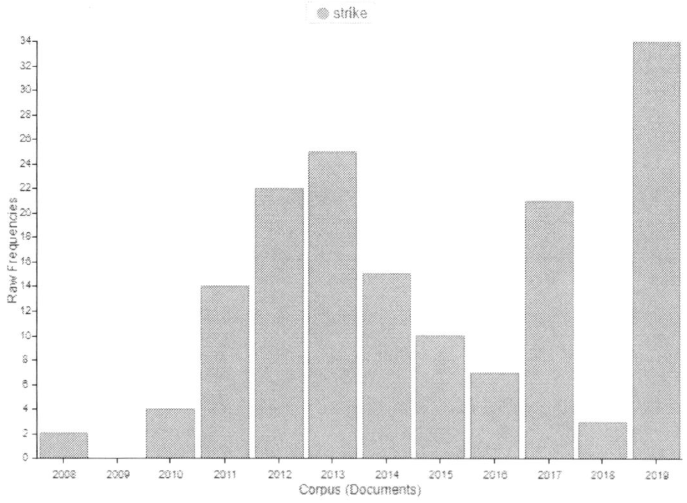

Figure 29. Column graph for the term strike used in the corpus of The Age from 2008–2019. ProQuest International Newsstream.

President Obama is mentioned 306 times, President Trump is mentioned 123 times, and President Bush is mentioned 38 times concerning 'drone strikes.' In 2019, President Trump was mentioned more times than any of the other presidents during any other year with 78 mentions (Figure 28). However, no US President is mentioned in 2018, not even President Trump, even though there were at least 1,985 drone strikes and at least 778 deaths in Afghanistan and 335 deaths from drone strikes in Somalia (Tables 1 and 2). The term 'strike' is mentioned 157 times and peaks in 2019 with 34 mentions. While the term drone appears more in 2012 and 2013, the term 'strike' appears more in 2019 most likely due to the increase in drone strike activity.

The Australian (Conservative)

Figure 30. Cirrus graph of the twenty-five most commonly used words in the corpus of The Australian from 2008–2019. ProQuest International Newsstream.

The Australian, a conservative newspaper published in New South Wales, Australia was founded in 1964 by Rupert Murdoch. It is owned and published by News Corp Australia, which is owned by Murdoch's international company News Corp. *The Australian* had 301 articles over the 12 years in our study, which had 321,048 total words and 17,564 unique words (Figure 30).

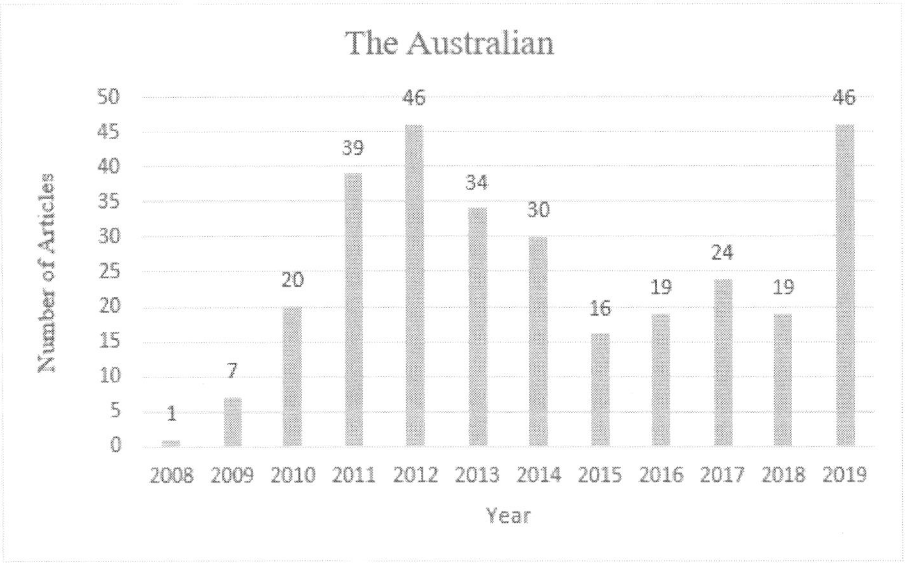

Figure 31. The number of total articles in The Australian for the years of 2008–2019 with the terms 'drone strike' and 'United States.' ProQuest International Newsstream.

Of the 301 articles found in *The Australian*, the articles peak in 2012 (and 2019) with 46 articles and then declined until 2015, which had only 16 articles (Figure 31). Over the next few years, there were slight increases and decreases, until 2019 where there were 46 articles, which is tied with 2012 as a peak year.

The Australian tended to have more positive articles in the beginning years and then many negative articles or neutral articles as the years passed. Overall, 22 positive articles were found, 42 negative articles, 22 neutral, and 22 had no reporting out of a total of 108 articles (Figure 32). During the Trump administration, one sees more no reporting articles, as they tended to discuss Russian or Iranian drones or Iran shooting down US drones. One example of a negative term that came up in this conservative newspaper was 'vaporized' to describe death by drone (Thomas 2018).

Figure 32. The Australian Content Analysis for 2008-2019. Coding in figure: P = Positive, N = Negative, NE = Neutral, and 0 = No Reporting. ProQuest International Newsstream.

	Australia - The Australian									
	Articles									
Year	1	2	3	4	5	6	7	8	9	10
2008 (1)	P									
2009 (7)	N	0	P	P	P	P	P			
2010 (20)	N	N	N	N	N	N	N	N	P	P
2011 (39)	N	N	P	P	P	P	N	N	P	P
2012 (46)	NE	N	P	P	N	N	NE	NE	NE	NE
2013 (34)	NE	NE	NE	N	N	NE	NE	NE	N	0
2014 (30)	N	P	N	N	N	P	P	N	NE	N
2015 (16)	NE	NE	NE	NE	N	N	NE	N	N	N
2016 (19)	NE	0	P	N	0	P	N	NE	N	N
2017 (24)	0	0	0	P	0	0	N	NE	0	N
2018 (19)	0	0	0	N	NE	NE	0	0	N	N
2019 (46)	0	0	N	0	0	N	N	0	0	0

	Term	Count
1	pakistan	739
2	military	729
3	drone	651
4	state	558
5	qa'ida	555
6	country	497
7	islamic	492
8	war	448
9	iran	446
10	attack	442
11	afghanistan	440
12	killed	432
13	syria	424
14	forces	421
15	general	413
16	strike	413
17	obama	402
18	defence	397
19	world	397
20	attacks	396
21	taliban	390
22	united	390
23	international	383
24	people	380
25	government	376

Figure 33. Top twenty-five most commonly used words in the corpus of The Australian. ProQuest International Newsstream.

The terms that appear the most in articles from *The Australian* are 'Pakistan,' 'military,' 'drone,' 'state,' and 'qa'ida.' 'Drone' makes the top 3 on the list, 'strike' is listed at number 16 (the plural 'strikes' does not make the top 25), 'killed' ranks in at 12 and 'war' is ranked number 8 (Figure 33). With *The Australian,* we see more country names and mentions with 'Pakistan' at the number 1 spot, 'Iran' at the number 9 spot, 'Afghanistan' at number 11, and 'Syria' at number 13. Unlike the newspaper *The Age*, we do not see mention of 'Iraq' and the mention of 'Iran' and 'Syria' in the top 25 is unique.

Distinctive words (compared to the rest of the corpus):

1. 2008: loudon (6), marriott (4), sadruddin (2), owais (2), inenglish (2).
2. 2009: hodge (25), bajaur (9), baxter (6), mehsud (17), displaced (12).
3. 2010: anonymous (58), headley (10), shahzad (14), hodge (24), tea (11).
4. 2011: anonymous (60), hodge (54), kashmiri (26), isi (32), haqqanis (23).
5. 2012: libi (39), nuremberg (14), hodge (48), romney (13), lamb (18).
6. 2013: sharif (48), blackwater (15), harnden (18), sharif's (17), hodge (42).
7. 2014: havard (43), abbott (37), maley (48), lyons (12), dna (27).
8. 2015: farag (25), prakash (29), ramadi (11), syria (88), jamie (10).
9. 2016: pine (29), prakash (37), gap (25), mansour (24), jabar (13).
10. 2017: abedi (35), kim (38), manchester (32), trump (47), korean (28).
11. 2018: israel's (29), israel (96), sofer (15), barak (13), australia's (16).
12. 2019: trump (185), iran's (64), pompeo (44), bolton (49), iran (298).

Figure 34. Most distinctive used words in the corpus of The Australian from 2008–2019. ProQuest International Newsstream.

The most distinctive words in the 301 articles in *The Australian* are different than what was seen in *The Age* (Figure 34). Unlike *The Age* or in the American newspapers, we do not see the terrorist organization 'ISIL' or 'ISIS' listed in the top 5 distinctive words in 2014. President Trump's name does appear in 2017, as seen in other papers, but so do other members of the Trump administration, such as United States Secretary of State Mike Pompeo and United States National Security Advisor John Bolton.

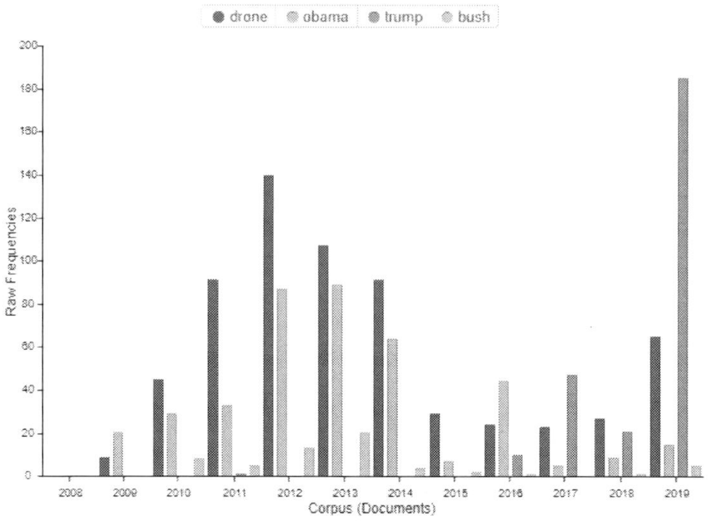

Figure 35. Trend graph on the raw frequency of the terms Bush, Obama, Trump, and drone used in the corpus of The Australian from 2008–2019. ProQuest International Newsstream.

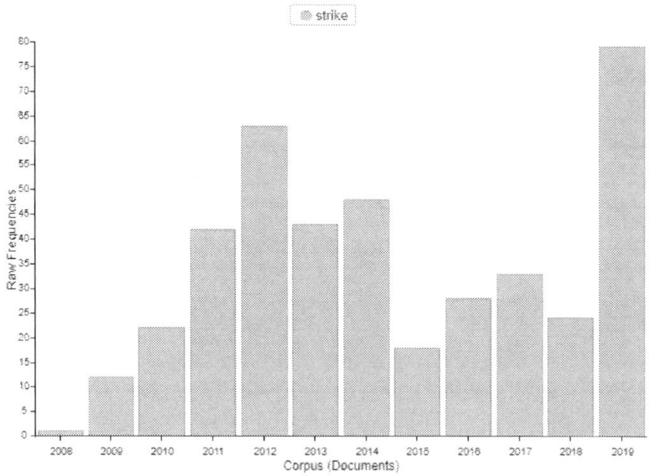

Figure 36. Column graph for the term strike used in the corpus of The Australian from 2008–2019. ProQuest International Newsstream.

President Obama is mentioned 402 times, President Trump is mentioned 264 times, and President Bush is mentioned 59 times with the term 'drone,' which is mentioned 651 times from 2008–2019 (Figure 35). The term 'strike' is mentioned 413 times and peaks in 2019 with 79 mentions (Figure 36).

Case 3 – Canada

Canada, the United States' neighbor to the north, has long been one of its closest political allies, but that does not mean that Canada and the United States always see eye-to-eye. Canada has six significant political parties, and many smaller ones, these include the Bloc Quebecois, Conservative Party of Canada, Green Party, Liberal Party of Canada, New Democratic Party, and the People's Party of Canada (CIA 2019). The Liberal Party who is center-left and the Conservative Party of Canada which is center-right are two dominant parties. Political power during the timeframe of this research from 2008–2019 included just two Prime Ministers: Stephen Harper of the Conservative Party of Canada (2006–2015) and Justin Trudeau (2015–current) of the Liberal Party of Canada (Parliament of Canada 2019) (Table 5).

Use of Canadian military drones and drone strikes

As of 2019, Canada did not have any armed drones, never used a drone in combat, and did not have any military drone strikes (New America 2019). However, in 2008 the Canadian Armed Forces flew leased Heron drones owned by Israel in Afghanistan for surveillance (Marijan 2020). In 2011, the Canadian Armed Forces requested funds to buy drones for use in the Libyan war, but their request was denied (Marijan 2020). The Canadian military continues to argue that drones will be of use for surveilling the Northwest passage, the Canadian Arctic, and other maritime areas as waterways become more navigable and popular (Marijan 2020).

Newspapers in Canada

As of 2020, there were 75 daily newspapers in Canada, which is down from 105 in 2015, but there are more than 900 community newspapers (News Media Canada 2020). Of the 75 titles, 71 titles are owned by corporations and 4 are independently owned (News Media Canada 2020). According to a 2021 report, most Canadians still prefer to read their newspapers in print, but Canada, like other places, is slowly losing its newspapers (Watson 2021). The Toronto Star is the largest newspaper in Canada, but *The Globe and Mail,* considered a liberal newspaper, and the conservative newspaper *The National Post* is known for its national and international coverage. A total of 176 articles were found in the *Globe and Mail* and 291 articles were found in the *National Post* under the study terms 'drone strike' and 'United States' for the years 2008–2019 (Figure 37).

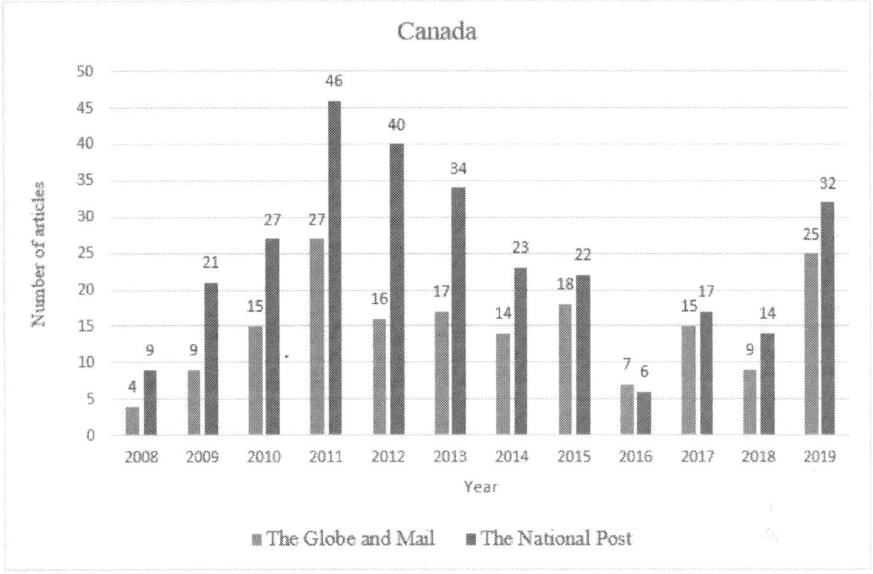

Figure 37. The number of total articles in the Globe and Mail and the National Post with the terms 'drone strike' and 'United States' years of 2008–2019. ProQuest Canadian Newsstream.

The Globe and Mail (Liberal)

Figure 38. Cirrus graph of the twenty-five most commonly used words in the corpus of the Globe and Mail for 2008–2019. ProQuest Canadian Newsstream.

The *Globe and Mail*, considered a liberal newspaper, is based out of Toronto, Canada, and is published by The Globe and Mail Inc. In 2019, the *Globe and Mail* had a daily circulation rate of 323,000 (World Atlas 2019). In the 12 years of this study, the *Globe and Mail* had 176 articles, which had a total of 239,687 words and 16,534 unique words (Figures 38 and 39).

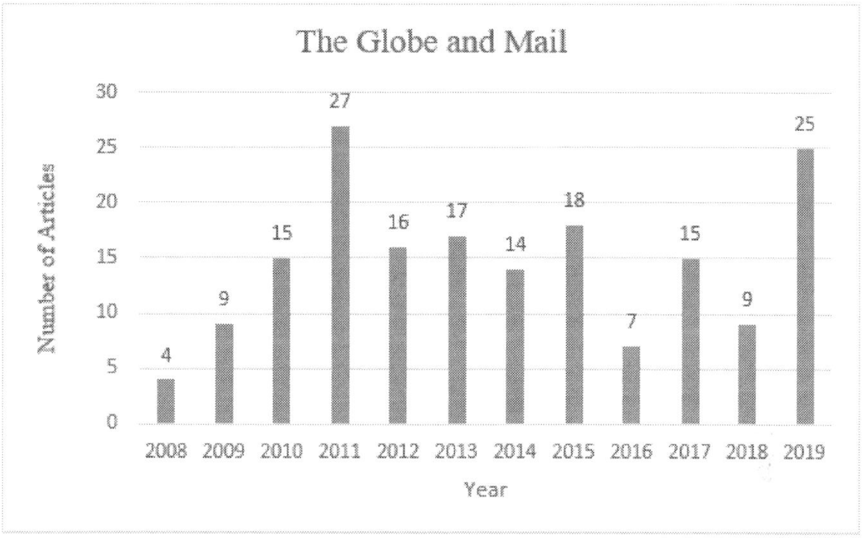

Figure 39. The number of total articles in the Globe and Mail for the years of 2008–2019 with the terms 'drone strike' and 'United States.' ProQuest Canadian Newsstream.

Of the 176 articles found for the *Globe and Mail*, the articles peak in 2011 with 27 articles and declined but stayed the same for the next four years until 2016 where only 7 articles were found (Figures 38 and 39). Then in 2017, the number of articles mentioning 'drone strikes' and the 'United States' increases, and in 2019, there were 25 articles.

The *Globe and Mail's* randomly picked articles show mostly negative articles or articles that have nothing to do with US drone strikes. There were 7 positive articles, 49 negative articles, 16 neutral articles, and 37 no reporting articles, out of a possible 109 articles (Figure 40). One example of a negative article comes from 2011 and says, "After nasty spats over gun-happy CIA agents killing Pakistani pedestrians, the humiliating American attack deep inside Pakistan that killed Osama bin Laden and a long-running duplicitous game over assassination strikes by U.S. Predator drones" (Koring 2011). The negative and neutral articles continue until 2017 when more articles had nothing to do with US drone strikes. What we see starting in 2017 are articles about Russian drones and Yemen's Houthi rebel drones as well as Iran shooting down a US drone.

Canada - The Globe and Mail

	Articles									
Year	1	2	3	4	5	6	7	8	9	10
2008 (4)	P	N	N	NE						
2009 (9)	NE	0	NE	N	N	N	NE	N	0	
2010 (15)	N	0	N	N	N	NE	N	N	N	NE
2011 (27)	N	N	N	0	N	0	N	N	N	0
2012 (16)	NE	P	N	0	0	0	N	P	N	N
2013 (17)	N	N	N	0	N	NE	N	N	NE	0
2014 (14)	N	N	0	N	P	NE	P	NE	P	NE
2015 (18)	0	NE	N	NE	N	N	N	N	N	N
2016 (7)	N	N	N	0	N	NE	N			
2017 (15)	N	NE	0	0	0	0	P	0	N	0
2018 (9)	N	0	0	0	N	0	0	0	0	
2019 (25)	0	0	0	0	0	0	0	0	0	0

Figure 40. The Globe and Mail Content Analysis for 2008–2019. Coding in figure: P = Positive, N = Negative, NE = Neutral, and 0 = No Reporting. ProQuest Canadian Newsstream.

	Term	Count
1	u.s	1233
2	military	503
3	war	493
4	united	487
5	pakistan	474
6	obama	464
7	states	439
8	canadian	411
9	drone	401
10	people	361
11	president	359
12	qaeda	358
13	country	355
14	strikes	340
15	government	337
16	afghanistan	327
17	killed	291
18	forces	272
19	general	270
20	years	267
21	attacks	262
22	air	261
23	attack	252
24	state	252
25	drones	247

Figure 41. Top twenty-five most commonly used words in the corpus of the Globe and Mail from 2008–2019. ProQuest Canadian Newsstream.

The terms that appear the most in articles from the *Globe and Mail* are 'US,' 'military,' 'war,' 'united,' 'Pakistan,' and 'Obama' (Figure 42). 'Drone' makes the top 9 on the list ('drones' comes in at number 25), 'strikes' is listed at number 14 (the singular 'strike' does not make the top 25), 'killed' ranks in at 17, and 'war' is ranked number 3. With the terms 'US,' 'military,' and 'war' in the top 3, and constantly used, it can be argued they give those articles a negative connotation.

Distinctive words (compared to the rest of the corpus):

1. 2008: perlez (5), morgan (5), gistaro (5), petraeus (8), yesterday (8).
2. 2009: bahadur (18), nazir (25), mcchrystal (13), mehsud (32), forks (8).
3. 2010: baradar (14), mullah (20), koring (35), torobin (6), roggio (6).
4. 2011: bouchard (15), bergen (15), pakay (12), koring (54), libyan (23).
5. 2012: pirates (17), mali (13), koring (29), romney (6), nizami (6).
6. 2013: mali (34), winchester (12), african (30), harvey (8), bryson (8).
7. 2014: wente (10), alexander (9), audience (10), bradshaw (7), dam (9).
8. 2015: aqap (22), globemedia (18), raqqa (13), schei (8), hessen (8).
9. 2016: chi (15), raq (13), decapitation (8), driver (17), strathroy (7).
10. 2017: sports (32), trump (50), whales (16), yar (38), slater (18).
11. 2018: hezbollah (61), israel (69), lebanese (18), debris (20), israel's (15).
12. 2019: uxo (57), dnd (38), energy (60), djibouti (31), iran (135).

Figure 42. Most distinctive words used in the corpus of the Globe and Mail form 2008–2019 ProQuest Canadian Newsstream.

The most distinctive words used in the entire selection of 176 articles give us a glimpse into unique words occurring when compared against the whole corpus (Figure 43). Terrorist organization 'ISIL' or 'ISIS' did not make an appearance in the most distinctive words, which has been seen in other newspapers. President Trump comes into the distinctive word used in 2017 and then is not seen again in 2018 or 2019. However, the terms 'Israel' and 'Hezbollah' do make an appearance in 2018 and 'Iran' in 2019.

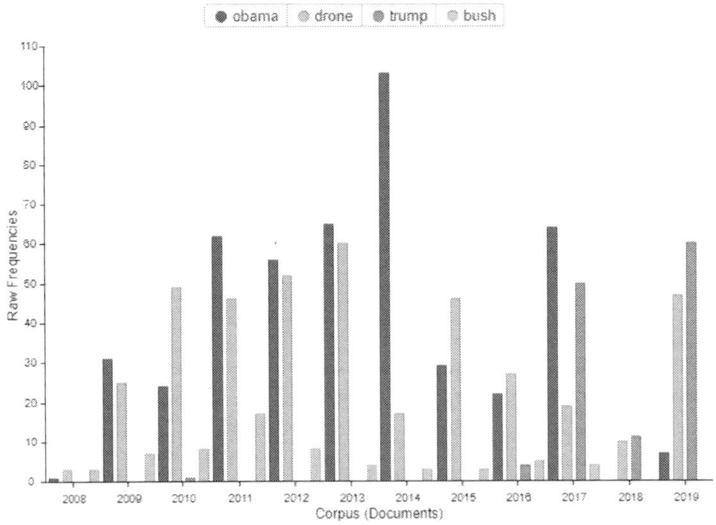

Figure 43. Trend graph on the raw frequency of the terms Bush, Obama, Trump, and drone used in the corpus of the Globe and Mail from 2008–2019. ProQuest Canadian Newsstream.

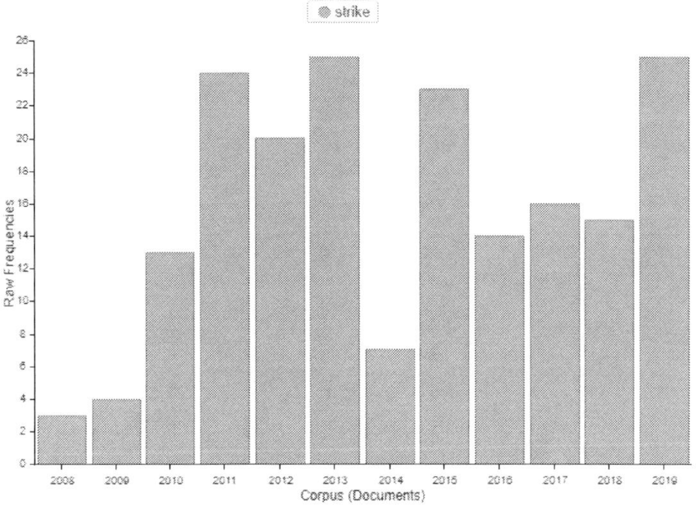

Figure 44. Column graph for the term strike used in the corpus of the Globe and Mail from 2008–2019. ProQuest Canadian Newsstream.

From 2008–2019, President Obama is mentioned in the *Globe and Mail* 464 times, President Trump is mentioned 126 times, and President Bush is mentioned 62 times, and the term 'drone' is mentioned 401 times. President Obama was mentioned more times, in 2014, than any of the other presidents during any other year with 103 mentions (Figure 43). This is different from other newspapers, such as the American and the Australian newspapers where President Trump is mentioned more in one year and that year tends to be 2019. The term 'strike' appears 189 times from 2008–2019, and the most mentions are in 2013 and 2019 with 25 mentions each year. The peak number of 'drone strike' articles in the *Globe and Mail* tie between 2013 and 2019. The term 'strike' appears 25 times in 2013 and 2019, but there were more articles in 2011 (Figure 44).

The National Post (Conservative)

Figure 45. Cirrus graph of the twenty-five most commonly used words in the corpus of the National Post from 2008–2019. ProQuest Canadian Newsstream.

The conservative newspaper, the *National Post,* is published by the Postmedia Network Inc. and is published in Don Mills, Ontario, Canada. The *National Post* had a daily circulation rate of 186,000 in 2019 (World Atlas 2019). In the 12 years of this study, the *National Post* has a total of 291 articles with the terms 'drone strike' and 'United States,' for a total of 350,800 total words and 20,553 unique words (Figure 45).

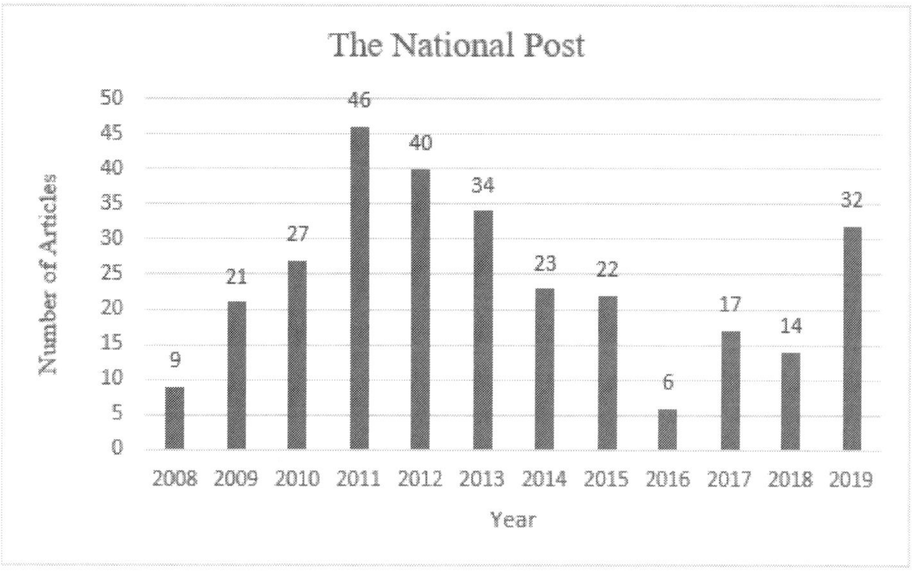

Figure 46. The total number of articles in the National Post for the years of 2008–2019 with the terms "drone strike" and "United States." ProQuest Canadian Newsstream.

Of the 291 articles found for the *National Post*, the articles peak in 2011 with 46 articles and gradually decline until 2016 where only 6 articles were found with the study's specific search terms (Figure 46). Then in 2017, the number of articles that mention drone strikes and the United States increase, and in 2019, there were 32 articles, numbers not seen since 2013.

Neutral and no reporting articles dominate in the *National Post* with 29 positive articles, 22 negative articles, 30 neutral articles, and 34 no reporting articles out of 120 possible articles (Figure 47). With positive articles, the *National Post* uses terms like 'insurgents' or 'militants.' The articles that came up as no reporting tended to be about Russian drones, Iran shooting down a US drone, and drones owned by terrorist organizations.

Canada - The National Post

Articles

Year	1	2	3	4	5	6	7	8	9	10
2008 (9)	P	N	0	0	0	P	NE	N	N	
2009 (21)	P	0	P	0	N	NE	P	0	P	P
2010 (27)	NE	NE	NE	P	NE	N	P	NE	P	NE
2011 (46)	0	N	N	P	N	N	N	N	P	NE
2012 (40)	P	0	P	P	P	0	0	P	NE	N
2013 (34)	NE	NE	NE	N	NE	P	0	N	P	0
2014 (23)	P	N	P	N	P	P	P	0	P	0
2015 (22)	P	NE	0	NE	P	0	0	0	0	NE
2016 (6)	0	NE	N	NE	NE	NE				
2017 (17)	NE	0	0	NE	NE	NE	N	NE	NE	0
2018 (14)	0	P	0	N	0	N	NE	NE	NE	P
2019 (32)	N	0	N	0	0	0	0	0	0	0

Figure 47. The National Post Content Analysis for 2008-2019. Coding in figure: P = Positive, N = Negative, NE = Neutral, and 0 = No Reporting. ProQuest Canadian Newsstream.

	Term	Count
1	u.s	1618
2	military	824
3	united	735
4	pakistan	710
5	states	663
6	war	620
7	drone	577
8	country	562
9	taliban	561
10	afghanistan	540
11	president	509
12	qaeda	500
13	obama	497
14	canadian	485
15	people	465
16	forces	459
17	government	436
18	iran	419
19	security	410
20	strikes	399
21	killed	396
22	attack	390
23	attacks	381
24	world	379
25	american	361

Figure 48. Figure Top twenty-five most commonly used words in the corpus of the National Post from 2008–2019. ProQuest Canadian Newsstream.

The terms that appear the most in articles from *The National Post* are 'US,' 'military,' 'war,' 'international,' 'Pakistan,' and 'states.' 'Drone' makes the top 7 on the list, 'strikes' is listed at number 20 (the singular 'strike' does not make the top 25), 'killed' ranks in at 21, and 'war' is ranked number 6 (Figure 48).

Distinctive words (compared to the rest of the corpus):

1. 2008: fusion (14), erie (13), tacit (9), dakota (7), goodspeed (18).
2. 2009: mehsud (44), mcchrystal (12), illustrations (12), qaeda (96), fledged (11).
3. 2010: gul (46), shahzad (31), wikileaks (28), cables (19), furlong (15).
4. 2011: gaddafi (54), gaddafi's (30), libyan (33), canwest (46), kashmiri (16).
5. 2012: romney (21), canwest (35), gaouette (12), qaeda (86), bowden (9).
6. 2013: sutherland (23), infomart (34), poachers (12), fisher (21), rhino (11).
7. 2014: isis (104), ukrainian (32), amerli (14), matt's (13), cbsa (13).
8. 2015: houthi (35), shuaib (18), houthis (30), isil (37), mma (13).
9. 2016: litvinenko (9), heuser (7), ramadi (9), hubbard (6), hauser (6).
10. 2017: trump (62), trump's (31), macisaac (11), executive (66), whitney (9).
11. 2018: farrow (16), idlib (14), ghouta (9), farrow's (9), humphreys (12).
12. 2019: trump (160), aramco (42), abqaiq (23), amazon (37), iran (169).

Figure 49. Most distinctive words used words in the corpus of the National Post from 2008–2019. ProQuest Canadian Newsstream.

The most distinctive words used in the entire selection of 291 articles give us a glimpse into unique words occurring when compared against the whole corpus. The terrorist organization ISIS makes an appearance in 2014. President Trump appears on the list in 2017 and is seen again in 2019 (Figure 49).

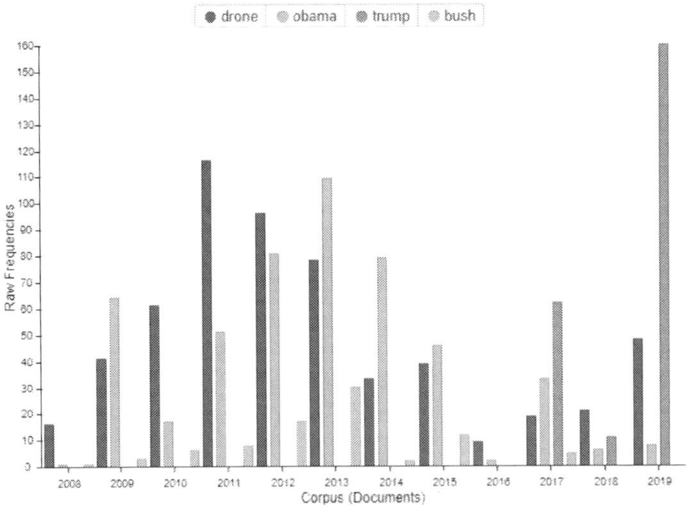

Figure 50. Trend graph on the raw frequency of the terms Bush, Obama, Trump, and drone used in the corpus of the National Post from 2008–2019. ProQuest Canadian Newsstream.

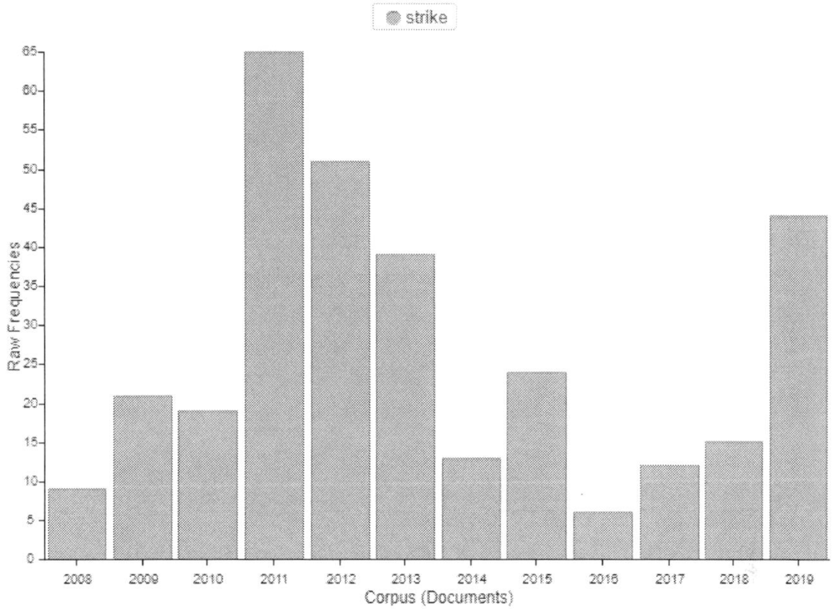

Figure 51. Column graph for the term strike used in the corpus of the National Post from 2008–2019. ProQuest Canadian Newsstream.

President Obama is mentioned 497 times, President Trump is mentioned 233 times, and President Bush is mentioned 84 times, and the term drone is mentioned 577 times from 2008–2019. President Trump was mentioned more times in 2019 than any of the other presidents during any other year with 160 mentions (Figure 50). The term 'strike' is mentioned a total of 318 times and peaks in the year 2011 with 65 mentions; 2012 has 51 mentions, and 2019 has 44 mentions. The peak number of 'strike' articles in the *National Post* fall in 2011 during the time period of the beginning of the US and NATO drawdown during the Obama administration (Figure 51).

Case 4 – New Zealand Newspapers

New Zealand has eight significant political parties and many smaller political parties, and they include the ACT New Zealand, Green Party, Mana Movement, Maori Party, New Zealand First Party, New Zealand Labour Party, New Zealand National Party, and the United Future New Zealand (CIA 2019). The National and Labour parties are the most popular, and the government of New Zealand during the timeframe of this research, 2008–2019, included four Prime Ministers: Helen Clark (1999–2008) of the New Zealand Labour Party, John Key (2008–2016) and Bill English (2016–2017) of the New Zealand National Party, and finally Jacinda Ardern (2017–2023) of the New Zealand Labour Party (New Zealand Government 2017) (Table 5).

Use of New Zealand military drones and drone strikes

As of 2019, New Zealand did not have any military armed drones, they never used a drone in combat, and did not launch any military drone strikes (New America 2019). However, the New Zealand Army has been training its soldiers to fly little civilian drones, such as the DJI Mavic Pro and DJI Phantom 4 (Bayer 2019).

Newspapers in New Zealand

The first newspaper in New Zealand was *The New Zealand Gazette* in 1839, but interestingly it was published in London and then shipped to New Zealand arriving in 1840, and then proceeded to change its name a few more times and eventually closed in 1844 (New Zealand Government 2020). In 1841, another newspaper called the *New Zealand Gazette* started publication and is now the official government newspaper (New Zealand Gazette 2021). Most of the newspapers in New Zealand are owned by two Australian companies – Australian Provincial Newspapers and the Fairfax Newspapers – but a few are owned by the Allied Press, and a small number are independently owned (Derby 2014). This study looks at *The Press* (liberal) and the *New Zealand Herald* (conservative) newspapers. In this study, a total of 85 articles were found in *The Press* and 170 articles were found in the *New Zealand Herald* under the terms 'drone strike' and 'United States' for the years 2008–2019 (Figure 52).

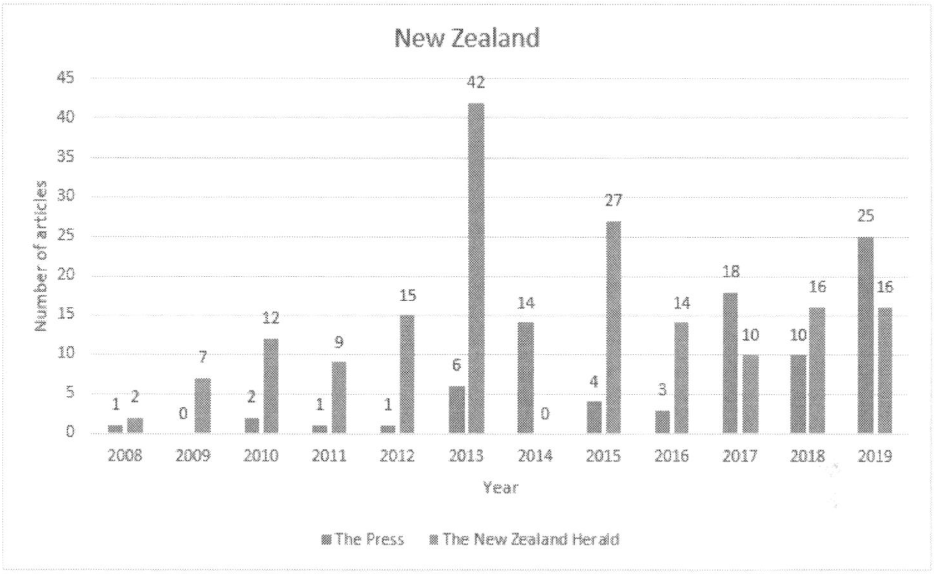

Figure 52. The total number of articles in The Press and The New Zealand Herald for the years of 2008–2019 with the terms 'drone strike' and 'United States.' ProQuest International Newsstream.

The Press (Liberal)

Figure 53. Cirrus graph of the twenty-five most commonly used words in the corpus of The Press for 2008–2019. ProQuest International Newsstream.

New Zealand's liberal or moderate newspaper *The Press* is based out of Christchurch, is owned by Stuff Limited, and is the oldest surviving newspaper on New Zealand's South Island (New Zealand Government 2016). Founded in 1861 by James FitzGerald, the first edition had just six pages (New Zealand Government 2016). In the 12 years of this study, *The Press* had a total of 85 articles for a total of 83,371 total words and 8,796 unique words (Figure 53).

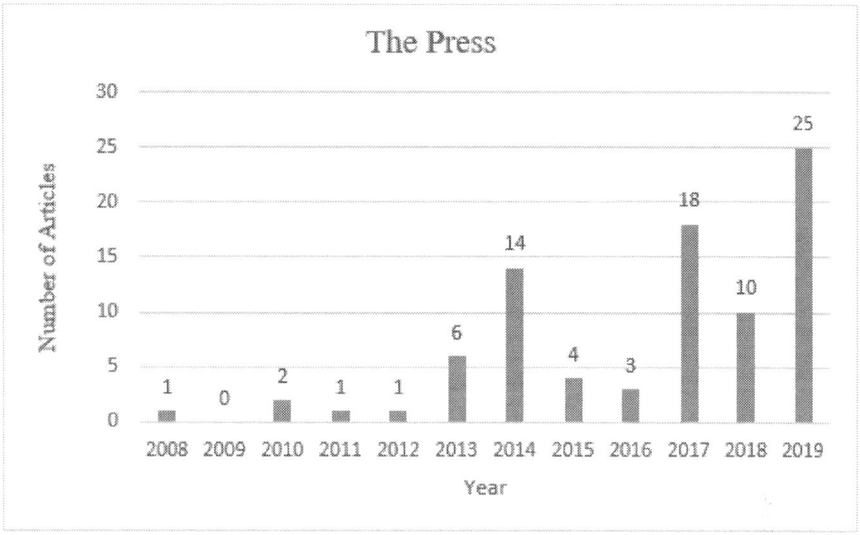

Figure 54. The total number of articles in The Press for the years of 2008–2019 with the terms 'drone strike' and 'United States.' ProQuest International Newsstream.

Of the 85 articles found for *The Press*, the articles peak at the end of the time frame of 2008–2019 as the first few years there are zero or very few articles from 2008 until 2012 (Figure 54). It is only in 2013 do we start to see some numbers and there is a spike in 2014 with 14 articles. A declinc is seen again in 2015 and 2016 and then in 2017, we see another huge spike in the numbers of articles that were written (Figure 54).

New Zealand - The Press

Articles

Year	1	2	3	4	5	6	7	8	9	10
2008 (1)	N									
2009 (0)										
2010 (2)	0	0								
2011 (1)	N									
2012 (1)	NE									
2013 (6)	N	N	NE	0	0	NE				
2014 (14)	N	NE	P	N	N	N	NE	NE	N	NE
2015 (4)	0	0	N	N						
2016 (3)	N	0	0							
2017 (18)	0	N	NE	N	N	NE	NE	0	NE	N
2018 (10)	N	N	0	0	0	0	N	P	N	0
2019 (25)	P	0	0	0	0	0	0	0	0	0

Figure 55. The Press Content Analysis for 2008–2019. Coding in figure: P = Positive, N = Negative, NE = Neutral, and 0 = No Reporting. ProQuest International Newsstream.

With 3 positive articles, 20 negative articles, 11 neutral, and 24 no reporting articles out of only 58 total articles, it's unsurprising that the majority of articles are negative and were found in 2014 and 2017, and also no reporting articles were very prominent during the Trump administration in 2018 and 2019 (Figure 55). The articles that are classified as neutral articles were articles that were written like news/press statements, there was no storytelling, and interestingly, some had a negative opinion or were criticizing the New Zealand government and Prime Minister and did not have a positive or negative view regarding the United States when the subject of drone strikes was discussed. Articles that had zero reporting turned up book reviews or discussions about Amazon's drone delivery program and Russian drones, but as these are not in the context of this study they are labeled zero reporting. A good example of a no reporting article from 2016 was about Russia and Syria, it mentions the US in passing and the term drone is about drone footage released by the Russian military leaving Aleppo (The Press, 2016).

	Term	Count
1	united	211
2	trump	192
3	drone	185
4	states	168
5	iran	166
6	international	158
7	military	146
8	nuclear	137
9	war	132
10	country	126
11	strike	118
12	killed	116
13	president	115
14	people	111
15	weapons	111
16	general	109
17	attack	106
18	forces	101
19	isis	98
20	state	96
21	yemen	96
22	drones	93
23	saudi	93
24	missile	92
25	russia	91

Figure 56. Top twenty-five most commonly used words in the corpus of The Press for 2008–2019. ProQuest International Newsstream.

The terms that appear the most in articles from *The Press* (Figure 56) are 'united,' 'Trump,' 'drone,' 'states,' and 'Iran.' 'Drone' makes the top 3 on the list ('drones' comes in at number 22), 'strike' is listed at number 11, 'killed' ranks in at 12 and 'war' is ranked number 9. With *The Press,* President Trump is shown in the top 25, something not seen in American, Canadian, and Australian newspapers. Also, of interest is the high ranking of 'Iran' at number 5, 'nuclear' at number 8, 'weapons' at 15, 'ISIS' at 19, 'Yemen' at 21, 'Saudi' at 23, and 'Russia' at number 25.

Distinctive words (compared to the rest of the corpus):

1. 2008: beynen (5), predator (11), horror (5), journal (3), robot (3).
2. 2009: none (1).
3. 2010: carell (17), hunting (15), fey (10), music (9), conservation (8).
4. 2011: afghan (9), taleban (5), imbalance (3), karzai (4), civilian (7).
5. 2012: mehsud (6), taleban (4), gravest (2), poses (2), pakistani (3).
6. 2013: gap (15), amazon (17), pine (16), xero (10), plot (9).
7. 2014: kobane (22), falls (20), havard (18), chris (13), yemen (34).
8. 2015: biosecurity (20), stink (12), smith (16), fruit (15), hersh (8).
9. 2016: aleppo (10), prakash (9), trump (17), egged (7), proposal (6).
10. 2017: isis (65), trump (41), mosul (15), raqqa (14), syria (37).
11. 2018: nuclear (43), putin (36), russia (32), iran (22), shabaab (9).
12. 2019: oil (72), saudi (92), iran (133), trump (122), pompeo (37).

Figure 57. Most distinctive words used words in the corpus of The Press from 2008–2019. ProQuest International Newsstream.

The most distinctive words used in the entire selection of 85 articles give us a glimpse into unique words occurring when compared against the whole corpus (Figure 57). There were no articles that had the terms 'drone strike' in 2009. Distinctive words from *The Press* include 'biosecurity' in 2015 and 'oil' in 2019.

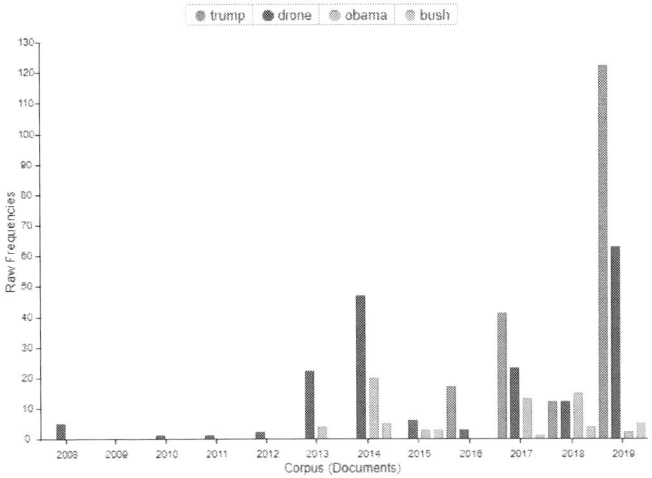

Figure 58. Trend graph on the raw frequency of the terms Bush, Obama, Trump, and drone used in the corpus of The Press from 2008–2019. ProQuest International Newsstream.

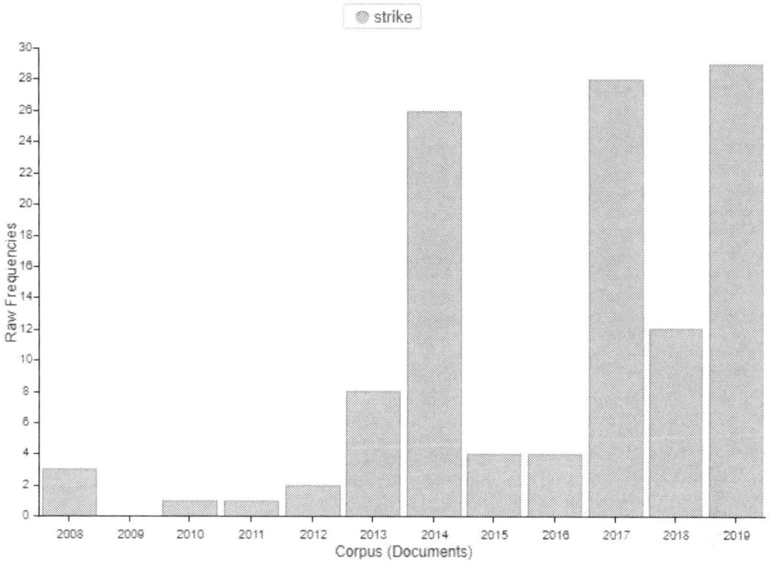

Figure 59. Column graph for the term strike used in the corpus of The Press from 2008–2019. ProQuest International Newsstream.

President Obama is mentioned 57 times, President Trump is mentioned 192 times, and President Bush is mentioned 18 times, and the term 'drone' is mentioned 185 times in 2008–2019. President Trump was mentioned more times in 2017 than any of the other presidents during any other year with 41 mentions (Figure 58). The term 'strike' hits its peak in 2019 with 29 mentions, 2017 has 28 mentions, and 2014 has 26 mentions. The peak number of 'drone strike' articles in *The Press* fall in 2019 during the Trump administration (Figure 59).

New Zealand Herald (Conservative)

Figure 60. Cirrus graph of the twenty-five most commonly used words in the corpus of the New Zealand Herald from 2008–2019. ProQuest International Newsstream.

The conservative newspaper the *New Zealand Herald* is based out of Auckland and is published by New Zealand Media and Entertainment. In the 12 years of this study, 170 articles were found with 181,365 total words and 14,072 unique words (Figure 60).

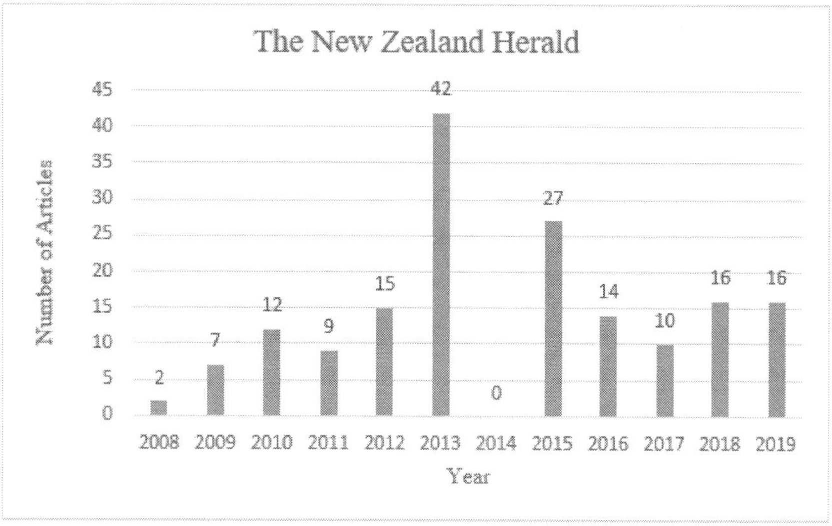

Figure 61. The number of articles in the New Zealand Herald for the years of 2008–2019 with the terms 'drone strike' and 'United States.' ProQuest International Newsstream.

The peak number of 'drone strike' articles in the *New Zealand Herald* fall in 2013 (Figure 61). There as the term 'strike' peaks in 2015 during the Taliban resurgence during the Obama administration. The year 2014 had no articles that fell under 'drone strike' and 'United States,' and multiple attempts were made in ProQuest International Newsstream and other ProQuest products. Considering there were articles for the years before and after 2014, one can conclude that the newspaper did not properly transfer their articles to the electronic sources or perhaps there were no articles for that year.

Looking at the content analysis of positive, negative, neutral, and no reporting on the articles in the *New Zealand Herald* find an interesting mixture in the early years, but then the negative articles become more prevalent with 30 negative articles, 20 positive, 18 neutral, and 29 no reporting articles out of a total of 97 possible articles (Figure 62). The newspaper used terms like counter-terrorism officials, militants, or missiles instead of the term 'drones' during the first years of the Obama administration, but then after 2015, those terms appear to stop. Interestingly, the articles use the term militants for positive stories of militants being killed by drones, but when they mention bin Laden, he is framed with the word terrorist. In years covering the Trump administration, the majority of articles fall under the category of 'no reporting,' and these include articles about the apparent drone assassination attempt on Venezuela's President Maduro, incidents involving New Zealand military personnel, usage of drones by Russian in Syria, Iranian drones shot down by Israel, Iran shooting down US drones, and Yemeni Houthi drones attacking Saudi Arabia oil fields.

New Zealand - The New Zealand Herald

Articles

Year	1	2	3	4	5	6	7	8	9	10
2008 (2)	P	NE								
2009 (7)	P	P	P	P	NE	P	0			
2010 (12)	NE	N	P	NE	N	N	NE	N	NE	P
2011 (9)	N	N	P	N	P	N	P	N	NE	
2012 (15)	P	NE	NE	P	P	NE	0	P	0	N
2013 (42)	N	N	N	N	N	N	N	P	NE	NE
2014 (0)										
2015 (27)	P	NE	NE	N	N	NE	0	N	P	0
2016 (14)	N	N	NE	N	0	N	NE	0	NE	N
2017 (10)	0	N	N	N	0	P	0	N	P	0
2018 (16)	0	0	0	0	0	0	0	0	N	0
2019 (16)	0	0	0	0	0	0	0	0	0	0

Figure 62. The New Zealand Herald Content Analysis for 2008–2019. Coding in figure: P = Positive, N = Negative, NE = Neutral, and 0 = No Reporting. ProQuest International Newsstream.

	Term	Count
1	drone	429
2	united	410
3	states	348
4	military	342
5	killed	308
6	attack	305
7	strikes	299
8	people	281
9	international	274
10	war	257
11	country	256
12	qaeda	253
13	security	241
14	intelligence	235
15	pakistan	230
16	obama	228
17	general	226
18	strike	224
19	world	217
20	president	216
21	government	214
22	iran	209
23	attacks	205
24	isis	205
25	drones	203

Figure 63. Top twenty-five most commonly used words in the corpus of the New Zealand Herald from 2008–2019. ProQuest International Newsstream.

The terms that appear the most in articles from the *New Zealand Herald* are 'drone' at the number 1 spot, and this is the only time in this study that the term 'drone' falls at the number 1 ranking. The term 'drone' is followed by 'united,' 'states,' 'military,' and 'killed' (Figure 63). 'Drones,' the plural term also falls on the list at 25, 'strikes' is listed at number 7 (the singular 'strike' at 18), 'killed' ranks in at 5, and 'war' is ranked number 10 (Figure 63). With the terms 'drone,' 'united,' 'states,' 'military,' and 'killed' at the top, these word choices come across as having a negative connotation.

Distinctive words (compared to the rest of the corpus):

1. 2008: lake (4), workplace (3), melamine (3), lottery (3), germs (3).
2. 2009: taleban (31), mehsud (39), baitullah (9), mehsud's (8), fat (6).
3. 2010: baghdad (57), gunshot (12), policeman (11), ied (9), baquba (9).
4. 2011: exhibition (8), album (6), abbot (6), taliban (18), uruzgan (5).
5. 2012: romney (44), cyber (30), laden (45), debates (8), manning (10).
6. 2013: brennan (25), nsa (15), records (13), briefs (18), guantanamo (23).
7. 2014: hits (1).
8. 2015: isis (107), jsoc (20), miller (25), counterterrorism (31), greg (17).
9. 2016: isis (63), propaganda (18), trump (28), mansour (17), jihadi (22).
10. 2017: trump (46), isis (27), chemical (18), tillerson (10), trump's (12).
11. 2018: maduro (22), google (25), ghouta (16), trump (31), israeli (44).
12. 2019: trump (69), prices (37), iran's (29), iran (118), zarif (19).

Figure 64. Most distinctive words used in the corpus of the New Zealand Herald from 2008–2019. ProQuest International Newsstream.

The most distinctive words used in the entire selection of 170 articles give us a glimpse into unique words occurring when compared against the whole corpus. The year 2014 had no articles that used the terms 'drone strike' and 'United States,' and 'hits' was just a random term that appeared in the generated list. We see the terrorist organization 'ISIS' listed in 2015, President Trump is seen in 2016, and even Guantanamo Bay is seen listed in 2013 (Figure 64).

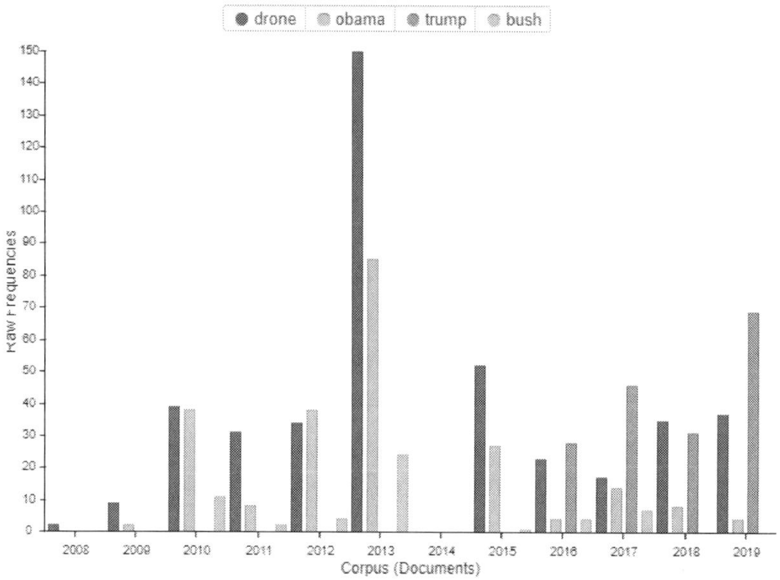

Figure 65. Trend graph on the raw frequency of the terms Bush, Obama, Trump, and drone used in the corpus of the New Zealand Herald from 2008–2019. ProQuest International Newsstream.

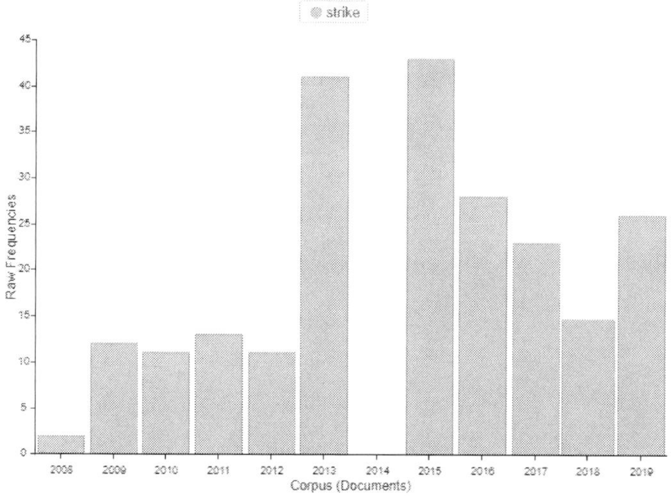

Figure 66. Column graph for the term strike used in the corpus of the New Zealand Herald from 2008–2019. ProQuest International Newsstream.

President Obama is mentioned 228 times, President Trump is mentioned 174 times, President Bush is mentioned 53 times, and the term 'drone' is mentioned 429 times from 2008–2019. President Obama was mentioned more times in 2015 than any of the other presidents during any other year with 85 mentions (Figure 65). With the term 'strike,' the peak year was 2015 with 43 mentions, 2013 had 41 mentions, and 2019 had 26 mentions (Figure 66). There were no articles found with the terms 'drone strike' in 2014.

Case 5 – The United Kingdom

The United Kingdom has at least twelve significant political parties and many smaller ones and they include the Alliance Party, the Conservative and Unionist Party, the Democratic Unionist Party, Green Party of England and Wales, Labour Party, Liberal Democrats, the Party of Wales, Scottish National Party, Sinn Fein, Social Democratic and Labor Party, Ulster Unionist Party, and the UK Independence Party (CIA 2019). The Conservative and Unionist Party, the Labour Party, and the Liberal Democrats have been the major parties in recent years, but recently, a coalition of two parties together to win power, which happened between 2010–2015. The government of the United Kingdom during the timeframe of this study, 2008–2019, consisted of Labour Party leader Gordon Brown (2007–2010), David Cameron (2010–2016) of the Conservative Party, Theresa May of the Conservative Party (2016–2019), and Boris Johnson (2019-2022) of the Conservative Party (Government of the United Kingdom 2019) (Table 5).

Use of United Kingdom military drones and drone strikes

The United Kingdom has used drones in military combat and first used an MQ-9 Reaper drone in Afghanistan in 2008 (New America 2019). They purchase their drones from the United States and Israel (New America 2019). In December of 2020, the British Army purchased 30 UAVs that are 'nano Bug drones' that can fit into the palm of your hand from the "UK weapons manufacture BAE Systems," and these small drones will allow the military to spy on targets up to 1.25 miles away (Jolly 2020). Then in 2020, the British Ministry of Defense purchased three Protector drones from General Atomics to replace their older Reaper drones for $80 million (Chuter 2020). Interestingly, British "aircrew flew approximately 2,150 missions using US Reaper and Predator drones, some of which were armed. The missions took off between October 2006 and December 31, 2012" and were embedded into the US "drone programme, where they have been flying each other's drones in Afghanistan, Libya and Iraq" (Savage 2013).

Newspapers in the United Kingdom

While many newspapers in the United Kingdom had their beginnings in the 1700s, it was not until 1836 that the growth of newspapers grew in popularity when the Stamp Tax was cut out (Bishop 1983, 206). While their number has decreased over the years, there are currently twelve daily newspapers in the United Kingdom, as well as eleven Sunday-only weekly newspapers. In this study, a popular liberal newspaper, *The Guardian,* and a conservative newspaper *The Times* were chosen. A total of 236 articles were found in *The Guardian* and 658 articles were found in *The Times* under the study terms 'drone strike' and 'United States' for the years 2008–2019 (Figure 67).

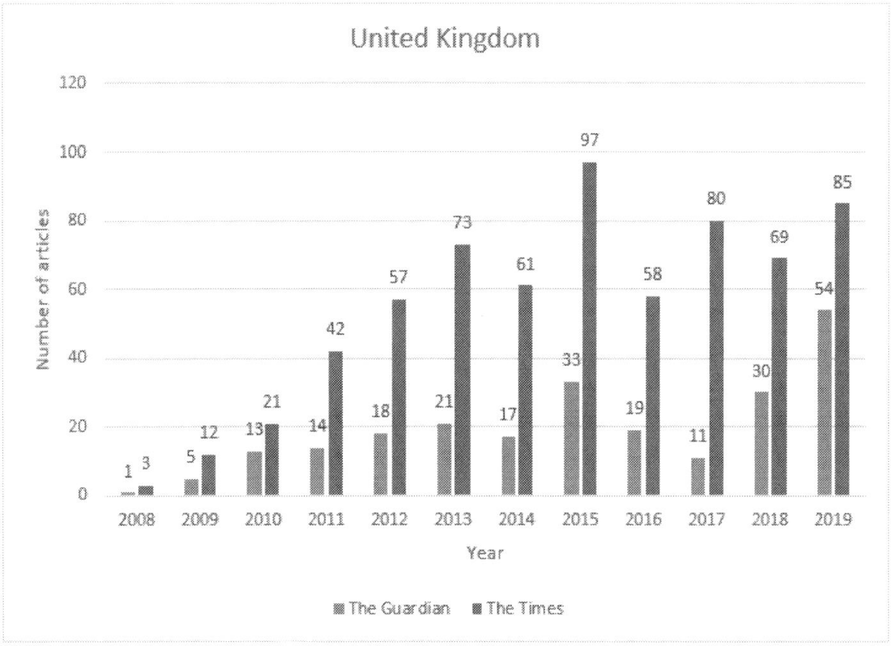

Figure 67. The number of total articles in The Guardian and The Times with the terms "drone strike" and "United States" years of 2008-2019. ProQuest International Newsstream.

The Guardian (Liberal)

Figure 68. Cirrus graph of the twenty-five most commonly used words in the corpus of The Guardian from 2008–2019. ProQuest International Newsstream.

The Guardian, a liberal newspaper in the United Kingdom, is based out of London and is published by the Guardian News and Media Limited, which is owned by the Scott Trust Limited. It was founded in 1821 and has a daily circulation rate in 2019 of about 134,000 (Statista 2019). In the 12 years of this study, *The Guardian* had 236 articles with 419,563 total words and 21,582 unique words (Figure 68).

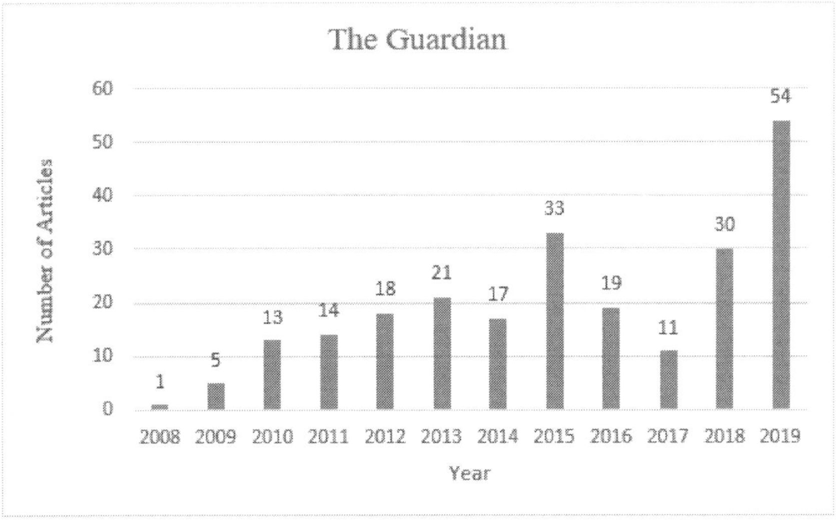

Figure 69. The total number of in The Guardian for the years of 2008–2019 with the terms "drone strike" and "United States." ProQuest International Newsstream.

Of the 236 articles found for the *Globe and Mail*, the articles peak at the end of 2019 with 54 articles. Starting in 2008, there is a gradual incline until 2013, then a decrease in 2014, only to have the numbers spike again in 2015 (Figure 69). Then after 2015, we see another decline until 2017 hits with 11 articles and then we see a rise in numbers in 2018 and that peak in numbers in 2019 with 54 articles.

The randomly chosen articles reviewed in *The Guardian* were predominately negative and some articles did not mention US drone strikes but were about drone strikes being launched by Russia, Israel, Iran, and the Royal Air Force (RAF). The majority of articles that mentioned drone strikes by the RAF were negative. Overall, only 12 positive articles were found, 35 negative articles, 15 neutral articles, and 44 articles had no reporting out of 106 possible articles (Figure 70). The last two years, 2018 and 2019, only had no reporting articles, so none of the articles used the search terms 'drone strike' and 'United States.'

	United Kingdom - The Guardian									
	Articles									
Year	1	2	3	4	5	6	7	8	9	10
2008 (1)	N									
2009 (5)	P	P	P	0	P					
2010 (13)	0	NE	N	NE	0	0	P	0	0	0
2011 (14)	0	NE	NE	N	NE	0	P	P	N	N
2012 (18)	P	N	N	NE	N	N	P	N	N	0
2013 (21)	N	N	N	0	N	N	0	0	NE	N
2014 (17)	0	N	NE	0	N	0	N	NE	N	NE
2015 (33)	N	NE	N	N	N	0	0	N	N	NE
2016 (19)	NE	P	0	0	N	0	P	0	0	N
2017 (11)	0	N	N	N	N	NE	P	N	N	NE
2018 (30)	0	0	0	0	0	0	0	0	0	0
2019 (54)	0	0	0	0	0	0	0	0	0	0

Figure 70. The Guardian Content Analysis for 2008-2019. Coding in figure: P = Positive, N = Negative, NE = Neutral, and 0 = No Reporting. ProQuest International Newsstream.

	Term	Count
1	united	751
2	people	731
3	war	729
4	military	704
5	isis	679
6	drone	650
7	obama	633
8	international	625
9	syria	618
10	country	577
11	government	544
12	iran	529
13	political	511
14	state	493
15	britain	472
16	iraq	471
17	british	459
18	trump	445
19	forces	439
20	saudi	436
21	world	433
22	qaida	426
23	attack	420
24	president	418
25	security	414

Figure 71. Top twenty-five most commonly used words in the corpus of The Guardian from 2008–2019. ProQuest International Newsstream.

The terms that appear the most in articles from *The Guardian* are 'united,' 'people,' 'war,' 'military,' and 'ISIS' (Figure 71). The term 'drone' makes the 6th spot on the list, 'strikes' does not make the top 25, 'killed' does not make the top 25, and 'war' is ranked number 3. This top 25 list is unique from other countries in that both 'Obama' and 'Trump' are listed, number 7 and 18, respectively. Both 'Britain' and 'British' are in the top 25, something not seen in other newspapers except *The Times*. Other countries that make the list are 'Syria,' 'Iran,' 'Iraq,' and 'Saudi.' The United States is probably signified under the term 'united' or 'states' or there could be a 'US' further down the list, as we have seen with other newspapers. More neutral terms like 'security' are used.

Distinctive words (compared to the rest of the corpus):

1. 2008: walsh (5), lahore (3), zooms (2), declan (3), stresses (2).
2. 2009: swat (15), baluchistan (13), mehsud (12), bahadur (8), gilani (11).
3. 2010: tf (19), 373 (26), cables (29), jpel (11), holbrooke (15).
4. 2011: laden's (33), gaddafi's (20), bin (105), tripoli (34), grief (9).
5. 2012: mcraven (14), seals (18), panetta (15), khar (9), bowden (9).
6. 2013: hashi (28), scahill (25), machiavelli (25), matrix (29), disposition (23).
7. 2014: welsh (72), wales (73), isis (87), wales's (21), devolution (21).
8. 2015: maqdisi (142), isis (402), qatada (70), baghdadi (53), syria (258).
9. 2016: submarines (108), trident (128), submarine (108), polaris (48), faslane (36).
10. 2017: trump (36), rohingya (10), trump's (13), suu (13), aung (13).
11. 2018: trump (110), trump's (27), holmes (25), syria (101), hamza (16).
12. 2019: trump (256), pompeo (57), bolton (80), trump's (54), iran's (49).

Figure 72. Most distinctive words used words in the corpus of The Guardian from 2008–2019. ProQuest International Newsstream.

The most distinctive words used in the entire selection of 236 articles give us a glimpse into unique words occurring when compared against the whole corpus (Figure 72). The terrorist organization 'ISIS' comes in 2014, President Trump comes in in 2017, and the term 'Wales' is mentioned numerous times. Also, in 2010, 'Task Force 373,' a joint military commando unit of coalition forces and other US Special Operations that used the term 'TF 373' to kill or capture suspected senior Taliban, but also civilian men, women, children, and Afghan police is mentioned.

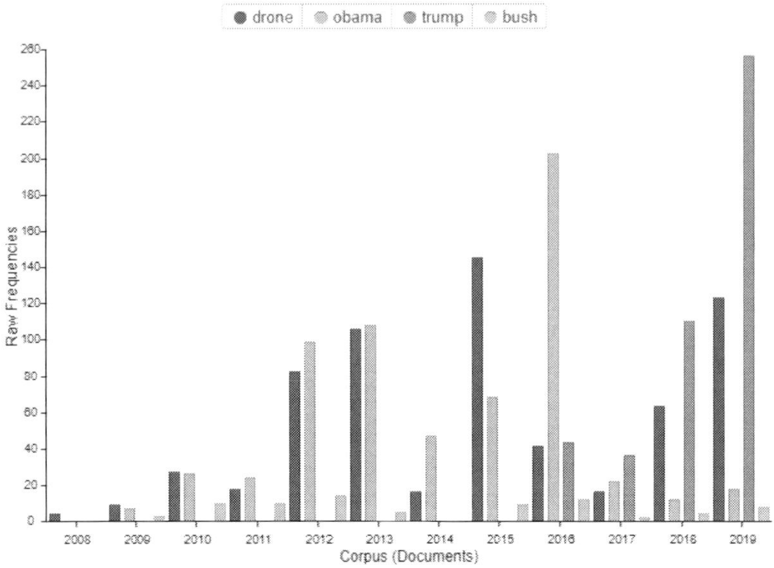

Figure 73. Trend graph on the raw frequency of the terms Bush, Obama, Trump, and drone used in the corpus of The Guardian from 2008–2019. ProQuest International Newsstream.

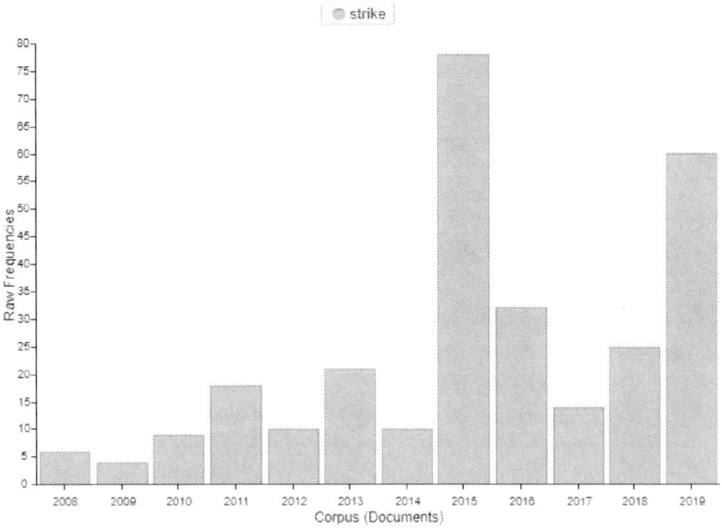

Figure 74. Column graph for the term strike used in the corpus of The Guardian from 2008–2019. ProQuest International Newsstream.

President Obama is mentioned 633 times, President Trump is mentioned 445 times, President Bush is mentioned 77 times, and 'drone' is mentioned 650 times in 2008–2019 (Figure 73). President Trump was mentioned more times in 2019 than any of the other US presidents during any other year with 256 mentions (Figure 74). None of the American presidents were mentioned in 2008, but 'drone strikes' were. The term 'strike' appeared 287 times with its peak of 78 times in 2015. Four years later in 2019, 'strike' would appear 60 times. The peak number of 'drone strike' articles in *The Guardian* fall in 2019 during the Trump administration, but interestingly, the term 'strike' appears more times in 2013 during the time period of the Taliban resurgence during the Obama administration.

The Times (Conservative)

Figure 75. Cirrus graph of the twenty-five most commonly used words in the corpus of The Times from 2008–2019. ProQuest International Newsstream.

The Times (also known as the *Times of London* or *The London Times*) was founded in 1785 and is considered a long-time conservative newspaper. Published out of London by the News International Trading Limited, its parent organization is News Corporation, owned by conservative Australian-American businessman Rupert Murdoch. It has a daily circulation rate of roughly 406,000 in 2019 (Statista 2019).

From 2008–2019 *The Times* has had a total of 658 articles, which then had 693,908 total words and 27,559 unique words (Figure 75).

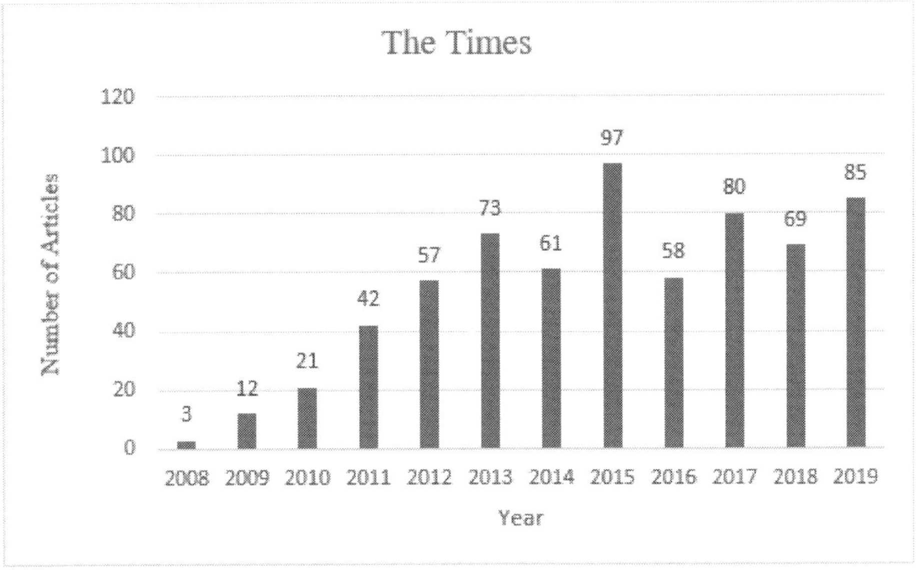

Figure 76. The number of total articles in The Times for the years of 2008–2019 with the terms 'drone strike' and 'United States.' ProQuest International Newsstream.

Of the 658 articles that were found for *The Times*, the drone strike articles peaked in 2015 with 97 articles with a gradual increase in the years leading up to 2015 and then a drop in 2016, only to have higher numbers again in 2017, 2018, and 2019 (Figure 76).

The Times presented a mix of 22 negative articles, 36 positive articles, 22 neutral articles, and 33 no reporting articles for a total of 133 articles (Figure 77). The articles that fell into the no reporting category included articles on drones belonging to Russia, Israel, Iran, and terrorist organizations as well as articles on drone strikes by the United Kingdom. Interestingly, one article revealed that 62% of the British public in 2014 supported RAF drone strikes against ISIS targets in Syria and that "the approval rating (61 percent) for drone strikes over Iraq" (Haynes 2014). Unlike *The Guardian*, *The Times* had articles that discussed US drone strikes in the last two years of 2018 and 2019.

	United Kingdom - The Times									
	Articles									
Year	1	2	3	4	5	6	7	8	9	10
2008 (3)	NE	N	P							
2009 (12)	0	P	N	N	NE	P	NE	N	NE	P
2010 (21)	0	N	P	0	N	P	P	NE	N	N
2011 (42)	P	NE	P	P	P	N	0	P	N	N
2012 (57)	N	P	N	N	N	N	P	N	0	P
2013 (73)	0	N	N	NE	P	N	0	NE	P	P
2014 (61)	P	P	P	P	0	NE	0	0	P	0
2015 (97)	0	NE	0	0	0	0	P	P	P	P
2016 (58)	NE	0	P	0	0	0	NE	N	NE	P
2017 (80)	NE	P	NE	P	P	P	P	0	0	NE
2018 (69)	0	0	NE	P	NE	0	0	P	0	0
2019 (85)	0	0	NE	N	NE	0	0	0	NE	NE

Figure 77. The Times Content Analysis for 2008–2019. Coding in figure: P = Positive, N = Negative, NE = Neutral, and 0 = No Reporting. ProQuest International Newsstream.

	Term	Count
1	united	1799
2	international	1511
3	drone	1491
4	isis	1327
5	military	1323
6	syria	1235
7	britain	1198
8	country	1048
9	general	1009
10	forces	981
11	killed	968
12	qaeda	941
13	pakistan	924
14	strikes	913
15	british	908
16	states	899
17	trading	896
18	kingdom	885
19	war	830
20	people	828
21	drones	821
22	attack	806
23	president	787
24	strike	775
25	iraq	765
26	government	759

Figure 78. Top twenty-five most commonly used words in the corpus of The Times from 2008–2019. ProQuest International Newsstream.

The terms that appear the most in articles from *The Times* are 'united,' 'international,' 'drone,' 'ISIS,' and 'military' (Figure 78). 'Drone' makes the top 5 on the list at number 3 ('drones,' the plural, comes in at number 21), 'strikes' is listed at number 14 (the singular 'strike' appears at 24), 'killed' ranks in at 11 and 'war' is ranked number 19. 'British' and 'Britain' also make the top 25 as do the terms 'Syria' and 'Iraq.' No US presidents make the top 25.

Distinctive words (compared to the rest of the corpus):

1. 2008: ha (11), ani (9), bajaur (12), taleban (28), jalaluddin (5).
2. 2009: taleban (73), mehsud (42), zainuddin (14), shamsi (24), gaza (34).
3. 2010: taleban (82), naseer (14), jsf (14), baradar (14), macintyre (14).
4. 2011: macintyre (37), giles (45), whittell (17), switchblade (9), lebor (9).
5. 2012: itty (36), aseefa (17), taleban (51), afridi (11), panetta (23).
6. 2013: bhat (41), hakimullah (32), superman (22), cw (22), mehsud (61).
7. 2014: isis (198), taylor (54), bhat (32), maliki (25), asiri (30).
8. 2015: isis (365), emwazi (105), syria (303), haynes (101), coghlan (98).
9. 2016: isis (242), hamilton (46), taliban (81), mansour (46), alm (15).
10. 2017: isis (266), abedi (44), spencer (116), trump (94), hamilton (62).
11. 2018: isis (152), trump (100), google (57), taliban (107), driverless (30).
12. 2019: trump (250), spencer (86), isis (104), bolton (38), rouhani (35).

Figure 79. Most distinctive words used in the corpus of The Times from 2008–2019. ProQuest International Newsstream.

The most distinctive words used in the entire selection of 658 articles give us a glimpse into unique words occurring when compared against the whole corpus (Figure 79). The terrorist organization 'ISIS' appears in 2014, President Trump appears in 2017 and continues in 2018 and 2019, as do some of his administration such as former National Security Advisor John Bolton.

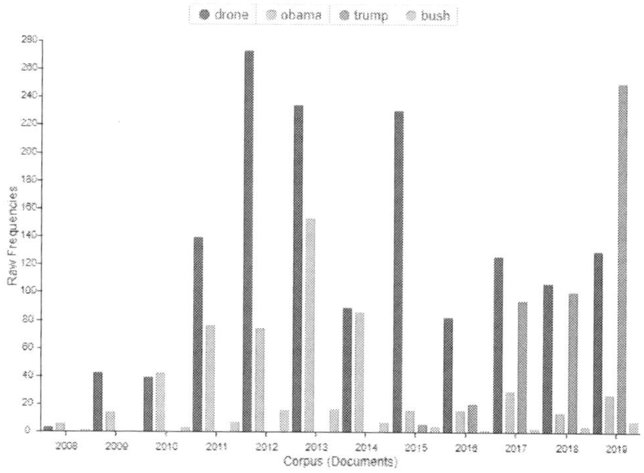

Figure 80. Trend graph on the raw frequency of the terms Bush, Obama, Trump, and drone used in the corpus of The Times from 2008–2019. ProQuest International Newsstream.

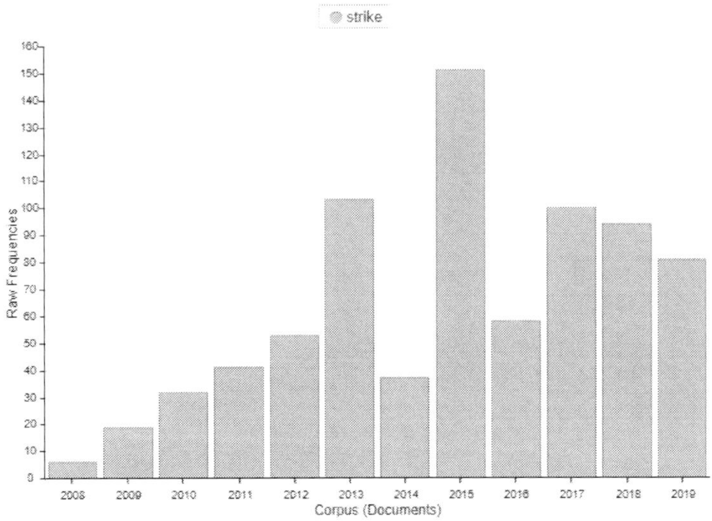

Figure 81. Column graph for the term strike used in the corpus of The Times from 2008–2019. ProQuest International Newsstream.

President Obama is mentioned 550 times, President Trump is mentioned 469 times, President Bush is mentioned 68 times, and the term 'drone' is mentioned 1,491 times from 2008–2019 (Figure 80). President Trump was mentioned more times in 2019 than any of the other presidents during any other year with 250 mentions. The term 'strike' peaks in 2015 with 151 mentions, but in 2019 there are only 81 mentions, so there is not an increase when there is an increase in drone strikes (Figure 81). The peak number of 'drone strike' articles in *The Times* fall in 2015 during the time period of the Taliban resurgence during the Obama administration. The term 'strike' also peaks in 2015 with the term appearing 151 times.

CHAPTER FIVE

CONCLUSION

In President Obama's 2020 autobiography he writes that "our national security agencies had been challenged to construct new forms of more targeted, nontraditional warfare – including operating an arsenal of lethal drones to take out al-Qaeda operations within the territory of Pakistan" (2020, 353). Based on the past three presidential administrations, it is safe to say that drones are now a regular feature of the US war chest and will continue to be part of any military strategy. This last chapter of the book concludes with the research summary and findings, the summary, the conclusion, limitations, contribution to the discipline, and further research.

Research summary and findings

Newspapers from the United States (the *New York Times* and the *Wall Street Journal*) had the largest number of articles on US drone strikes, 1,700 and 909 articles respectively, as one would expect as these are US drones (Table 6 and Figures 82, 83, and 84). United States newspapers, regardless of their political leanings, liberal or conservative, were writing about drone strikes more in the early years of this study than they were in later years, with the peak years between 2011 to 2015 (Figures 82, 83, and 84). In 2010 and 2011 during President Obama's troop surge, the number of drone strike articles by both the *New York Times* and the *Wall Street Journal* were not at their highest, so major events are not a cause for an increase in newspaper articles. The most dramatic example not seen in the data was in 2019 when over 7,167 drone strikes were unleashed in Afghanistan, up from 1,985 in 2018 and 2,609 in 2017, signaling a new policy by the Trump administration to "intensify pressure on the Taliban and force them into a settled negotiation to end the war" (Snow 2020) (Table 6). The Trump administration's new policy of bombing Afghanistan was to force the Taliban to settle the peace negotiations showed a slight rise in drone strike articles; they were not the numbers that were seen earlier in such years as 2012 or 2013.

When looking at the 76 articles in 2019 that fit the parameters of this study for the *Wall Street Journal* and including the term Afghanistan, none of the articles mentioned the 7,000 plus drone strikes. The 10 articles also did not discuss US drone strikes, as the drone strikes mentioned were from Israel and Iran (Figure 85). What was surprising was that the *New York Times* was found in the 10 random articles to have more positive than negative articles. This is interesting in that one would expect the liberal newspaper to have more negative articles (Figure 85).

It would be safe to speculate that international newspapers would have noticed and opposed more than 7,000 plus US drone strikes in 2019, but that does not seem to be the case. Drone strikes articles are not seen in the data, so the US's English-speaking allies were not writing about it in their newspapers. Without having the numbers of drone strikes from other sources, one would not know that they were happening (or falling as the case may be). The *New York Times* reported in September of 2019 that the "number of American airdropped missiles and bombs in Afghanistan is set to meet or possibly outpace the 7,362 munitions launched in all of 2018," so the US and international press were aware of what was happening (Zucchino 2019). Overall, the international newspapers told another story and that was one mainly of not agreeing with the United States and its drone strike policy (Figures 82, 83, and 84). These international news articles also told the story of how towards later years during the Trump administration any mention of drone strikes in Afghanistan and Iraq stopped. For the most part, US drone strikes were seen in a negative light. Drone strikes from other state actors, such as Russia and Iran, and even Israel was also seen as negative, as were drones from terrorist organizations. Interestingly the few mentions of US drone strikes in Libya were seen in a positive light. Australia and Canada have only surveillance drones and they tend to receive a mixed message in their press. The United Kingdom has had drone strikes that have killed people, and they also receive a negative view from the newspapers and the public, but the UK military is very excited about purchasing drones and having them in their arsenal (Figure 85).

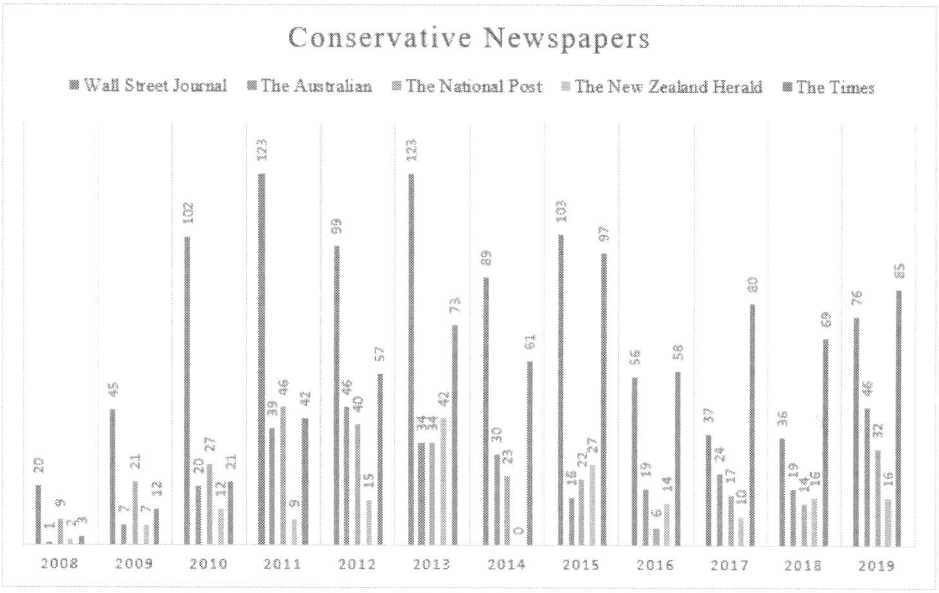

Figure 82. Number of conservative newspaper articles for the years of 2008–2019 with the terms 'drone strike' and 'United States.' ProQuest Canadian Newsstream, ProQuest International Newsstream, and ProQuest US Newsstream.

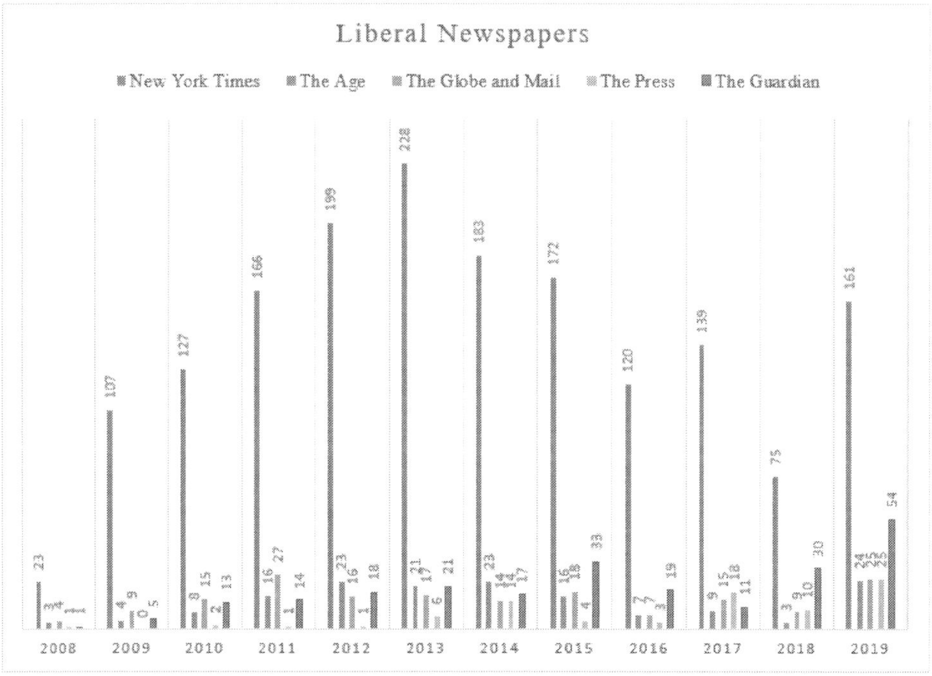

Figure 83. Number of conservative newspaper articles for the years of 2008–2019 with the terms 'drone strike' and 'United States.' ProQuest Canadian Newsstream, ProQuest International Newsstream, and ProQuest US Newsstream.

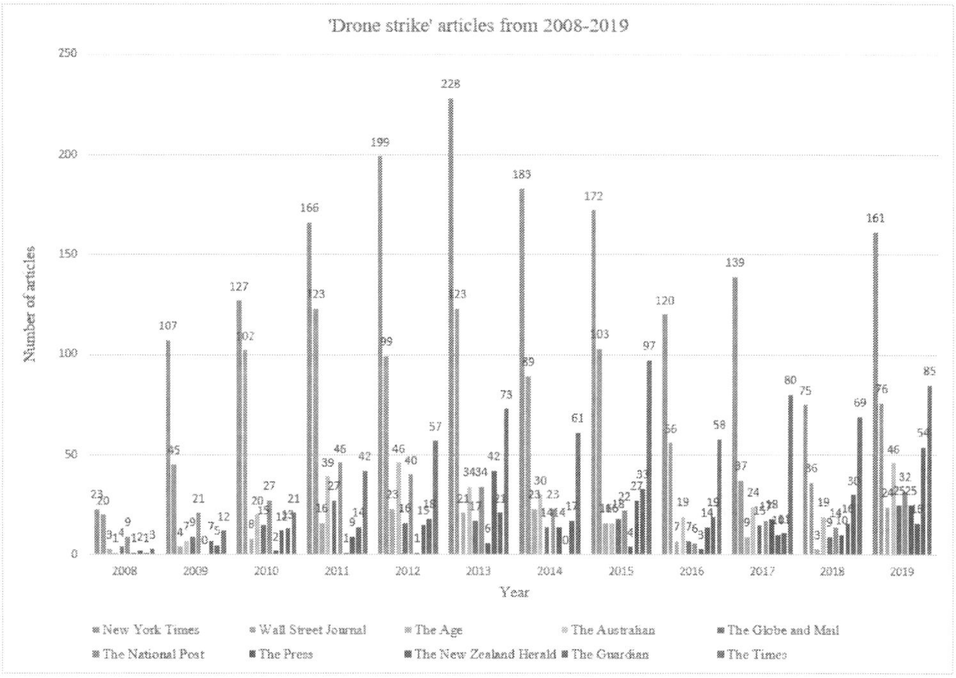

Figure 84. The total number of drone articles for the years of 2008–2019 with the terms 'drone strike' and 'United States.' ProQuest Canadian Newsstream. ProQuest International Newsstream, and ProQuest US Newsstream.

	P = Positive	N = Negative	NE = Neutral	0 = No Reporting	Total	
New York Times	43	22	21	34	**120**	0
Wall Street Journal	54	17	15	34	**120**	0
The Age	6	31	21	36	**94**	26
The Australian	22	42	22	22	**108**	12
The Globe and Mail	7	49	16	37	**109**	11
The National Post	29	22	30	34	**115**	5
The Press	3	20	11	24	**58**	62
The New Zealand Herald	20	30	18	29	**97**	22
The Guardian	12	35	15	44	**106**	14
The Times	36	22	22	33	**113**	7

Figure 85. Breakdown of positive, negative, neutral, and no reporting articles used in this study. ProQuest Canadian Newsstream, ProQuest International Newsstream, and ProQuest US Newsstream.

Both Australian newspapers (*The Age* and *The Australian*) were found to emphasize more negative articles (Figure 85). Many articles also fell into the neutral and no reporting categories. By 2012, *The Age* describes what is happening, "while support for the war in Afghanistan has fallen to an all-time low in all NATO states, the drone war is popular in the US. That's hardly surprising, given it offers no danger to American forces – the ultimate asymmetric warfare – while supposedly 'taking out' terrorists. But these high-tech death squads are creating a dangerous global precedent, which will do nothing for US security" (Milne 2012). While it was expected that the liberal *The Age* would have more negative articles, the fact that *The Australian*, the conservative newspaper, had as many negative articles is interesting. In both newspapers towards the final years of this study 2017, 2018, and 2019, they both tend to have more no reporting articles than positive or negative or neutral. So if they were discussing drone strikes they were not American drone strikes.

Our neighbors to the north, the Canadian newspapers (The *Globe and Mail* and The *National Post*) (Figure 85) were split, with the *Globe and Mail* emphasizing the negative articles and the *National Post* with more positive articles. Neutral articles were also emphasized in the *National Post* articles. Both had more no reporting articles in 2019 in the 10 randomly picked articles, so newspapers in Canada did not seem to be discussing American drone strikes (Figure 85).

The New Zealand newspapers (*The Press* and the *New Zealand Herald*) emphasized mostly a negative viewpoint or were not discussing the topic of US drone strikes (Figure 85). *The Press* had few mentions of US drone strikes and the majority of them were negative. The conservative newspaper, the *New Zealand Herald,* in 2012 mentioned how the US and Australia were considering putting a joint drone base on the Crocs Keeling Islands to use the drones to patrol the western Pacific and the Indian Ocean. While the *New Zealand Herald* does have more positive articles than *The Press*, that does not make them pro-Trump and in the end, they had more negative articles than positive. One article in 2017 begins with the sentence "Everyone take a deep breath and calm down. For weeks now, every newspaper and news channel is full of the latest sins of the Great Satan, Donald Trump. He's the instant bogeyman, suddenly emerging from the fiery pit, leaving the commentariat all aquiver in fear and trepidation" (Rudman 2017). Both newspapers had more no reporting articles in 2018 and 2019 among the 10 randomly picked articles, so once again newspapers did not seem to be discussing American drone strikes.

Finally, newspapers from the United Kingdom (*The Guardian* and *The Times*) were also split with more negative articles in *The Guardian*, and more positive articles in *The Times* (Figure 85). Both newspapers also had a high number of neutral newspapers. Of the 10 randomly picked articles, *The Guardian* had no reporting articles for the last two years, 2018 and 2019, while *The Times* had neutral and no reporting articles for 2019. While one would expect more positive articles from the conservative *The Times* newspaper, there were many more negative and neutral articles than was expected.

The overall finding of this study shows that no newspapers reported on US drone strikes equally with the same amount of newspaper articles, and in reality, the number of articles is lower as the search parameters of this study also picked up drones and drone strikes that were not American, such as those by Russian, Iranian, and British. These were classified in the study as no reporting. While the *New York Times* was the newspaper that reported on drone strikes with the highest frequency each year, even they were not writing about them in 2019 as they had in previous years, such as 2012 and 2013.

Interestingly, presidential approval ratings (Figure 1) do not correlate with drone strikes, because if it did then we would see more drone strikes in 2009 at the high of President Obama's popularity and see less of them with the decline of President Trump's approval ratings and that is not the case.

One wonders if the drop off in domestic and foreign newspaper coverage of US drone strikes at the end of the Obama administration and during the Trump administration was a 'donor fatigue' situation. This author speculates that there are two 'donor fatigue' situations occurring during the time frame of this study from 2008 to 2019. The first one involves the War on Terror. After almost 20 years of war in Afghanistan, Iraq, and Pakistan, newspapers, their editors, reporters, and their readers are just tired of reading about the war, and they are unknowingly experiencing a 'donor fatigue' situation. Newspapers, regardless of liberal or conservative views, are fatigued on reporting on drone strikes and the death they cause, and unless it's a big event, readers are tired of reading about it. News on the war and drone strikes does not sell. The second part of 'donor fatigue' started when President Trump became president in 2017. International newspapers, as seen by the quote in the *New Zealand Herald*, were concerned in the very beginning about President Trump and his administration, and by 2019, newspapers, the journalists, their editors, and their readers around the world were tired of American tweets and drama. Because most in the United States were also exhausted from four years of political fighting from both major parties, when the latest round of drone strikes started in Afghanistan in 2019, with numbers never seen before, there was no public outcry, and no one seemed to be interested. If the American or international press as well as the American public were interested, it would have been more prevalent in the numbers for articles about drone strikes, because newspapers write about popular topics (topics that will sell newspapers and advertising) and this is not seen in the data for American newspapers, and it is not seen in newspapers for our English-speaking allies. What is seen is that they moved on to other topics, other issues, other conflicts, and other drone strikes.

Summary

In my analysis, this author found that not one format told the whole story about the data – the articles on US drone strikes. Reading the 10 randomly picked articles told one part of the story, and all 4,683 articles entered into *Voyant Tools* told a different story. They both gave you different views and clues as to what the whole drone story might be but did not tell you enough of a full picture on their own. Context is everything.

Often the data is there, but if you blink you could miss it. For example, when looking at the *Voyant Tools* data for the US conservative newspaper the *Wall Street Journal* usage of the terms 'drone,' 'strikes' are lower than in the liberal newspaper the *New York Times*, so they are used less frequently. Uniquely terms such as 'war' and 'killed' do not make the top 25 in the conservative newspaper, and a unique term such as 'permission' is commonly used enough to make the top 25 when it ranks at number 15. When a newspaper is not using negative words like 'war,' 'killed,' and using positive terms like 'permission' with the words 'drone strikes' it illustrates an authorization for their usage and shows that the *Wall Street Journal* has a more supportive attitude toward drone strikes.

What was interesting about this study was how even with just 10 randomly picked articles, one could still see the long arc of a very long war and how drone and drone strikes were appearing in the literature and public consciousness, and we see them coming into use in Pakistan, Afghanistan, Somalia, and Yemen and later we see them in Libya. Drone strikes have become part of the conflict in Libya and for the most part receive positive press from the international newspapers, even those that are anti-US drone strikes, which goes to show that it matters what country is getting bombed and what the situation is. In 2008, there were only three countries using drones to kill: Israel, the United States, and the United Kingdom, but at the end of the study in 2019, we see the arrival of Russian drones bombing Syria, and we also see Iran backing Yemen's Houthi rebels with drones so that they can target and destroy the Saudi Arabian oil facilities (Wassef 2021). We view other historical events through this data, such as the death of bin Laden, and politicians – such as presidents and secretaries of state come and go on the international scene, we see how countries approve, don't approve, and ignore the US and its military policies, such as in Pakistan and its love/hate relationship with the US and US drones, and we see different events, like Iran shooting down American drones.

Closing Observations

The results of this study were unique. The first question asked: Has American newspaper coverage of US drone strikes remained positive from 2008–2019?

What was seen in US newspapers, the *New York Times* and the *Wall Street Journal,* is that they remain positive towards US drone strikes throughout the years of this study, as both newspapers had more positive articles than negative articles in this study in the 10 articles that were randomly picked. While the number of positive articles might not have increased as the years went on, they did hold steady. The liberal newspaper, the *New York Times,* wrote more about drone strikes than any other newspaper in this study, almost double that of its American conservative competitor the *Wall Street Journal*. However, even the *New York Times* did not show a higher frequency of articles when drone strikes increased during the Trump administration. With over 7,167 drone strikes in Afghanistan in 2019 and an increase in drone strike deaths in Somalia in 2018 (335 deaths) and 2019 (326 deaths), drone strike articles did not dramatically increase and are on par for the 2015 numbers. For the most part, as the number of drone strikes increased, the US newspapers remained positive.

The second question asked: Has coverage opposing US drone strikes increased in the English-speaking newspaper of US English-speaking allies from 2008–2019?

Opposition to drone strikes did not increase in the English-speaking allies' newspapers as the number of US drone strikes increased. There were more negative articles in the middle years of this study, during the Obama administration, and not more when one would expect them because of the increase in drone strikes, which totaled over 7,000 strikes, in 2019 during the Trump administration. The conservative newspapers of our English-speaking allies, *The Australian*, the *National Post*, and the *New Zealand Herald* all had more negative articles than positive or neutral articles towards the final years. *The Times* had more neutral articles towards the end of the time period of this study. The liberal newspapers seemed to stop reporting on drone strikes towards the last few years of this study and have more no reporting articles, which means that they were reporting on other drone strike topics than US drone strikes. In general, after the number of articles peak in 2014 and 2015, their numbers decreased in 2016, 2017, and 2018, with a small bump in numbers in 2019, but nowhere near the numbers seen in 2014 or 2015. So as the number of US drone strikes was going up in 2019, the opposition to them was not growing or going up, because if there was more opposition with our English-speaking allies', newspapers would see an increase in the data with more newspaper articles and with 7,167 US drone strikes in 2019 we do not see this.

The final question asked: Do the political leanings of newspapers influence the narrative for support for or opposition to US military drone strikes from 2008-2019?

The political leanings of English-speaking allies' newspapers do tend to influence a positive or negative view of US military drone strikes, but it is not a one size fits all situation. In general, conservative English-speaking allies' newspapers are more positive toward US drone strikes, and the liberal English-speaking allies' newspapers are more negative towards US drone strikes. For example, the conservative British newspaper *The Times* had a few more positive articles toward the end of the study than did *The Australian*, but there were more negative articles, neutral or no reporting articles in the conservative newspapers as the years went on and especially during the Trump administration. As the years went on, the liberal newspapers had more negative articles and then had more no reporting articles, so they just tended to stop writing about US drone strikes. Discussion and the number of articles on US drone strikes in both conservative and liberal newspapers dwindled and overall did not increase as the years go on with 2019 being the exception and even 2019 numbers were not what they once were. In the American newspapers, one sees positive and no reporting articles in the liberal and conservative newspapers as the years go on and even during the Trump administration, which is very different from our English-speaking allies' newspapers.

Overall Findings

Overall, this study found that conservative newspapers were different from each other, even though most are owned by the same corporation, News Corp, which is owned by conservative Australian-American businessman Rupert Murdoch. There would seem to be some form of independence of the topic or just cultural differences shown through in the newspaper articles. For example, the *Wall Street Journal* was more positive on US drone strikes, but then *The Australian* was not. Australia did have a liberal government for the last half of the study and so did Canada, but national leadership does not seem to have been a factor in how newspapers viewed US drone strikes. Out of a random selection, the *National Post* of Canada had more positive articles, but it had almost as large numbers of neutral and negative articles too, so one could easily classify that newspaper as more neutral, especially when compared to the large number of negative articles found in the *Globe and Mail*. The liberal newspapers, except for the *New York Times*, were all found to have more negative articles, but they did not increase in time, and in fact, their articles on US drone strikes dried up after the Obama administration ended. If their opposition was to increase one would expect to see more newspapers opposing the Trump Administration 7,000+ drone strikes and that is not seen in the data.

This study is easily understood when you combine soft power and hard power to create smart power. Soft power is easily seen in the influence of the United States around the world, especially in the years following the end of the Cold War, the American public's approval of anything military in the years following 9/11, the US military's continual explanation that drone strikes are precise and that they minimized the risk to soldiers and reduced collateral damage around civilians. Soft power can also be seen in the decline and opposition to the US military and its drones by citizens around the world and our English-speaking allies. Hard power is easily seen in the might of the US military as well as the drones themselves with increased firepower. So, in the end, you need smart power to explain the use of US military drone strikes and why the United States English-speaking allies' governments continue to support drones as seen in their newspaper articles.

Limitations

Limitations of this research include access to newspapers from all countries and databases that included all the articles. If a researcher does not have access, such as via a university subscription, the only other option is to purchase a personal subscription to each newspaper, and this can be expensive. Also, word choices and search parameters can create limitations in a study's parameters. Finding keywords that will cast a wide net on the topic is essential. Limitations were also found in picking only 10 random articles per year per newspaper to examine up close. On one hand, they provided up close and personal reports and stories about drone strikes, but on the other hand, they were random and so another sample of articles might have produced other results. Finally, any database is only as good as what it holds. For example, the year 2014 for the *New Zealand Herald* was missing, and multiple searches at different times never uncovered any articles in the ProQuest databases that were used.

Besides data, limitations are also found in the form of technology. While *Voyant Tools* has been around for a few years and it is open-source software and free to the public, there is no guarantee that it will be there from one day to the next. *Voyant Tools* does have a version of the software that can be downloaded to a researcher's desktop, but if the platform disappears there would be no way to update or fix any software issues that might appear.

Finally, finding drone strike data and drone strike death data is not an easy undertaking. The US government will not release its drone strike data (occasionally some limited information is released), so the only option is to use data from various global NGOs, and they all collect data with slightly different parameters. They are constantly improving and updating their data and then they release it to the public. The drone strike data is not 100% complete, as they are working after the fact or after the bombs have dropped, but until the US government releases its numbers it is the best source of information available.

Further Research

The role of the drone in American warfare is just in its infancy, and as technology improves with smaller, faster, cheaper, and more deadly drones, its role is just beginning. Shaw argues that the "US military is at the beginning of an ontology of war that shifts power projection from discrete platforms (such as expensive fighter jets and aircraft carriers) to amorphous and autonomous swarms" (2017, 459). What this means is that drone studies are in their early stages and present a growing field of research for scholars.

For the US military, drones are about power, and so this will interest a lot of future scholars. In 2011, the German newspaper *Der Spiegel* wrote that "the future of warfare belongs to drones...they allow a military force to exert power while minimizing its own risks, and to carry out precise, deadly strikes, without sending its own soldiers into danger. The additional fact that drones are comparatively cheap has made them a favorite with the United States" (Lorenz, von Mittelstaedt, and Schmitz 2011). Smart power helps explain the use of military drone strikes around the world by the United States and why the United States English-speaking allies governments continue to support drone strikes even though the global public does not. Drones and drone strikes need smart power to work and be accepted or acceptable, which leads to an interesting question of what happens when a nation or individual does not have smart power and uses drones to kill. Sharkey writes that "currently drones are used with ease against low-tech communities. More technologically sophisticated opponents would adopt counter strategies such as jamming satellite signals to render them useless or bring them down. A fully autonomous drone could still seek out its target without human intervention" (2012, 26). Soon, the US might be faced with this question: what happens when enemy nations use their drone strikes on US soldiers or US territory or when our English-speaking allies use them? This opens new avenues of study that will move beyond how drones are currently being studied and interpreted.

Afterward

This study only examined the years of 2008–2019 and a lot has changed since 2019 and when this book was published in early 2024. A new US President, Democrat Joseph Biden took office on January 20, 2021, and on August 30, 2021, he ended the War in

Afghanistan. In 2022 President Biden also signed a new "classified policy tightening the rules for the CIA and the Pentagon to carry out deadly drone strikes, and commando raids conducted outside of traditional war zones" (Lillis 2022). US drones continue to be used for counterterrorism strikes in Afghanistan, Somalia, and Yemen (Lillis 2022; Sterman 2022). Drone usage in war also seems to be entering a new era with both Ukraine and Russia using drones in the war in the Ukraine, with Russia using military drones mainly purchased from Iran, the Ukraine using civilian drones, and with the new war between Israel and Hamas (Myre 2023). So, the more things change, the more they stay the same.

APPENDIX A

Publicly Available Data

This study uses data and information that is publicly available online from the company websites. Data was collected from these newspapers on July 24, 25, 26, and 27, 2020:

The New York Times, United States of America
The Wall Street Journal, United States of America
The Age, Australian
The Australian, Australian
The Globe and Mail, Canada
National Post, Canada
The Press, New Zealand
The New Zealand Herald, New Zealand
The Guardian, United Kingdom
The Times, United Kingdom

Other drone data was gathered from:

The Bureau of Investigative Journalism
The New America Foundation
The Long War Journal
Amnesty International
Human Rights Watch
Stanford/NYU Law

ACKNOWLEDGMENTS

There are so many people to thank and acknowledge that it is hard to know where to begin. This book is the result of my Ph.D. research and dissertation from the University of Southern Mississippi. Love and thanks to my parents Carole and Terry Aho, who have always been supportive of whatever crazy idea or thing I have decided to do. To my twin sister Michelle Aho, I love you, but you now have to call me Doctor Melissa. Thanks to the four-legged feline members of the family: Che Aho, Candy Aho, and Ginger Aho. Thank you to my friends and family: Sloane, Derek, Louisa, and Julian Kosky, Kelley and Marc Boyer, Susan Wakefield, Dr. Cece Churchill, Jane Pollari and Ray Holmgren, Steve and Sue Gulczinski, Kathryn and Tom Truax, Janet and Dave Ranney, Josh Truax, Randy Kopitzke, and Dan Peterson.

As I have been saying for years, it takes a village to raise a Ph.D. student. Thank you to my IDV Ph.D. friends and family. First of all, thank you to Dr. Joy Patton and Dr. Madeline Messick whom I met on the first day of our Ph.D. program. My crazy map still works, because "hello" we lived and survived the curse of Shaggy's! Thanks also go out to Dr. Michelle Watts, Brian Messick, Dr. Issam Abu-Ghallous, Dr. Todd Barry, Mary Jane Barry, Sean Barry, Dr. Keith Newcomer, the late Dr. Dave Davis, Dr. Heather Martin, Dr. Kevin Frank, and Pepe Soto. Thank you also to their wonderful families whom I have been fortunate to get to know over the years.

Thank you to my long-suffering Ph.D. dissertation chair Dr. Joseph J. St. Marie who listened to my 101 crazy dissertation ideas and was incredibly supportive while I was trying to finish my Ph.D. in the middle of a global pandemic. Sadly, Dr. St. Marie, I could never come up with that zombie dissertation the world is waiting for. Many thanks to my Ph.D. committee members Dr. Robert Pauly, Dr. Tom Lansford, and Dr. Michael Wigginton. Thank you to Annette Copeland, Graduate Coordinator extraordinaire of the International Development program at the University of Southern Mississippi, who has been so kind and helpful to me and other Ph.D. students over the years.

REFERENCES

Alexander, David. 2013. "Retired General Cautions Against Overuse of "Hated" Drones." *Reuters,* January 7, 2013. https://www.reuters.com/article/us-usaafghanistan-mcchrystal/retired-general-cautions-against-overuse-of-hated-drones idUSBRE90608O20130107/

Arkin, Ronald. 2009. *Governing Lethal Behavior in Autonomous Roberts*. Boca Raton, Florida:Chapman and Hall/CRC.

Atherton, Kelsey D. 2013. "Flying Robots 101: Everything You Need to Know About Drones." *Popular Science*, March 7, 2013. https://www.popsci.com/technology/article/2013-03/drone any-other-name

Bayer, Kurt. 2018. "NZDF has No Plans to Ground Drones Banned by US Military Allies Over Cyber-Safety Fears." *New Zealand Herald*, March 1, 2018. https://www.nzherald.co.nz/nz/nzdf-has-no-plans-to-ground-drones-banned byus-military-allies-over-cybersafetyfears/M5YWKT57O3OBVI434IVSJMSHNE/

Bennett, W. Lance. 2016. *News: The Politics of Illusion, Tenth Edition*. 10th edition. Chicago, Illinois: University of Chicago Press.

Berg, Bruce L. 2009. *Qualitative Research Methods for the Social Sciences*. 7th edition. Boston,Massachusetts: Pearson.

Bishop, Robert L. 1983. "The Decline of National Newspapers in the UK." *Gazette (Leiden, Netherlands),* 31 (3): 205–212. https://doi.org/10.1177/001654928303100305/

Blaszczak-Boxe, Agata. 2014. "Drone Pilots Suffer PTSD Just Like Those in Combat." *Livescience.com*, August 20, 2014. https://www.livescience.com/47475-drone-operators-develop-ptsd.html

Boyle, Michael J. 2020. *The Drone Age: How Drone Technology Will Change War and Peace.* New York City, New York: Oxford University Press.

The Bureau of Investigative Journalism. 2019. "Drone Warfare." Accessed May 6, 2019. https://www.thebureauinvestigates.com/projects/drone-war/

_____. 2021. "Drone Wars: The Full Data." Accessed April 6, 2021. https://www.thebureauinvestigates.com/stories/2017 01-01/drone-wars-the-full data/

Byman, Daniel L. 2013. "Why Drones Work: The Case for Washington's Weapon of Choice." *Brookings* (blog), June 17, 2013. https://www.brookings.edu/articles/whydrones-work-the-case-for washingtons-weapon-of-choice/

Campbell, Raelene C. 2018. *Tweets About Tornado Warnings: A Spatiotemporal and ContentAnalysis.* MS, United States – South Carolina: University of South Carolina.

Cavelty, Myriam Dunn, Sophie-Charlotte Fischer, and Thierry Balzacq. 2016. "'Killer Robots' and Preventive Arms Control." In *Routledge Handbook of Security Studies*, edited by Myriam Dunn Cavelty and Thierry Balzacq, 2nd edition, 457-468. London, United Kingdom: Routledge.

Chuter, Andrew. 2020. "UK Orders First Three Protector Drones from General Atomics." *Defense News,* July 15, 2020. https://www.defensenews.com/global/128anada/2020/07/15/uk-orders-first-three-protector-drones-from-general-atomics/

CIA (Central Intelligence Agency). 2019. "Political Parties and Leaders." *The World Factbook.* 2019. https://www.cia.gov/library/publications/the-world factbook/fields/315.html/

Clausen, Christian. 2017. "Air Force to Retire MQ-1 Predator Drone, Transition to MQ 9 Reaper." *U.S. Department of Defense,* February 27, 2017. https://www.defense.gov/Newsroom/News/Article/Article/1095612/air-force to-retire-mq-1-predator-drone-transition-to-mq-9-reaper/

Cohen, Grant M. 2018. *Origins of the U.S. Public Opinion for Drone Strikes: The Intersection of Elite Rhetoric, Media Coverage, and American Public Opinion, 2000-2015.* Ph.D., United States – Florida: University of Miami. https://scholarlyrepository.miami.edu/oa_dissertations/2068

Cortes-Sanchez, Julian David. 2018. "Mission Statements of Universities Worldwide Text Mining and Visualization." *Intangible Capital* 14 (4): 584-603.

Custers, Bart. 2016. "Drones Here, There and Everywhere Introduction and Overview." In *The Future of Drone Use: Opportunities and Threats from Ethical and Legal Perspectives*, edited by Bart Custers, 1st ed. 2016 edition, 3-20. New York City, New York: T.M.C. Asser Press.

Daines, J. Gordon III, Cory L. Nimer, and Jacob R. Lee. 2018. "Exploring the American Archivist: Corpus analysis tools and the professional literature." *Journal of Contemporary ArchivalStudies* 5 (article 3).

Derby, Mark. 2014. "Ownership of Major New Zealand Newspapers, 2014). *Teara – the Encyclopedia of New Zealand*. October 22, 2014. Accessed April 20, 2021. https://teara.govt.nz/en/map/43196/ownership-of-major-new-zealand newspapers-2014

Drake, Bruce. 2013. "Report Questions Drone Use, Widely Unpopular Globally, but Not in the US" *Pew Research Center* (blog). October 23, 2013. https://www.pewresearch.org/fact-tank/2013/10/23/report-questions-drone-use widely unpopular-globally-but-not-in-the-u-s/

Dobriansky, Paula. 2013. "Forward." In *Smart Power: Between Diplomacy and War*, Christian Whiton, ix-xiii. Washington, D.C.: Potomac Books.

The Economist. 1994. "Save the Front Page." *The Economist*, September 3, 1994. The Economist Historical Archive.

Encyclopaedia Britannica. 2019a. "Australian Labor Party." https://www.britannica.com/topic/Australian-Labor-Party

______. 2019b. "Democratic Party." https://www.britannica.com/topic/Democratic/Party/

______. 2019c. "Liberal Party of Australia." https://www.britannica.com/topic/Liberal/Party-of-Australia/

Freeze, Colin, 2015. "A Radacation-Free Conversation." *The Globe and Mail*. April 22, 2015: A6. ProQuest Canadian Newsstream.

Gao, Wenli. 2017. "Text Analysis of Communication Faculty Publications to Identify Research Trends and Interest." *Behavioral & Social Sciences Librarian* 36 (1): 36 47. https://doi.org/10.1080/01639269.2017.1507223.

George, Alexander L., and Andrew Bennett. 2005. *Case Studies and Theory Development in the Social Sciences*. Fourth Printing edition. Cambridge, Massachusetts: The MIT Press.

Gordon, Michael R. 2013. "Kerry, in Pakistan, Expresses Optimism on Ending Drone Strikes Soon." *New York Times,* August 2, 2013: A.6. ProQuest US Newsstream.

Government of the United Kingdom. 2019. "Past Prime Ministers." The United Kingdom. May 2019. https://www.gov.uk/government/history/past-prime-ministers

Graham, Shawn, Ian Milligan, and Scott Weingart. 2016. *Exploring Big Historical Data: The Historian's Macroscope.* London, United Kingdom: Imperial College Press.

Grayson, Kyle. 2016. *Cultural Politics of Targeted Killing: On Drones, Counter Insurgency, and Violence.* London, United Kingdom: Routledge.

Hagan, Joe. 2005. "Circulation Continues to Decline at Most Major Newspapers." *The Wall Street Journal, Eastern Edition,* November 8, 2005.

Hagey, Keach, Lukas I. Alpert, and Yaryna Serkez. 2019. "Local News Fades Out: A Divide between Newspaper Haves and Have-Nots." *Wall Street Journal, Eastern Edition,* May 4,2019.

Haynes, Deborah. 2014. "Support for Airstrikes on Isis in Syria is Growing." *The Times*. October 23, 2014:14. *ProQuest International Newsstream*.

Herman, Edward S., and Noam Chomsky. 2002. *Manufacturing Consent: The Political Economy of the Mass Media*. New York City, New York: Pantheon Books.

Insinna, Valerie. 2020. "Boeing Rolls out Australia's First 'Loyal Wingman' Combat Drone.'*Defense News,* May 4, 2020. https://www.defensenews.com/air/2020/05/04/boeing-rolls-out-australias-first loyal-wingman-combat-drone/

Jha, U. C. 2014. *Drone Wars: Ethical, Legal and Strategic Implications*. New Delhi, India: KW Publishers Private Limited.

Jolly, Jasper. 2020. "UK Army Buys 30 'Bug' Drones That Can Spy on Targets 2km Away." *The Guardian,* December 27, 2020. https://www.theguardian.com/world/2020/dec/28/uk-army buys-30-bug drones-that-can-spy-on-targets-2km-away/

Kreps, Sarah E. 2016. *Drones: What Everyone Needs to Know*. New York City, New York: Oxford University Press.

Koring, Paul. 2011. "U.S.-Pakistan Relations Sink to New Low. *The Globe and Mail.* July 9, 2011. ProQuest Canadian Newsstream.

Krishnan, Armin. 2009. *Killer Robots: Legality and Ethicality of Autonomous Weapons.* Farnham, United Kingdom: Ashgate.

Lillis, Katie Bo. 2022. "Biden finalizes New Rules for US Drone Strikes." *CNN Politics,* October 7, 2022. Accessed June 21, 2023. https://www.cnn.com/2022/10/07/politics/drone-strikes-counterterrorism white-house-biden-new-rules/index.html

Lorenz, Andreas, Juliane von Mittelstaedt, and Gregor Peter Schmitz. 2011. "'Messengers of Death': Are Drones Creating a New Global Arms Race?" *Der Spiegel* & *ABC News*, October 21, 2011. Accessed February 14, 2021. https://abcnews.go.com/International/drones-creating-global-arms-race/story?id=14788147/

Marald, Erland, Nancy Langston, Anna Stens, and Jon Moen. 2016. "Changing Ideas in Forestry: A Comparison of Concepts in Swedish and American Forestry Journals During the Early Twentieth and Twenty-First Centuries." *Ambio* 45 (Suppl 2): s74 s86. Doi: 10.1007/s13280-015-0744-7/

Maramba Inocencio Daniel, Antoinette Davey, Marc N. Elliott, Martin Roberts, Martin Roland, Finlay Brown, Jenni Burt, Olga Boiko, and John Campbell. 2015. "Web based textual analysis of free-text patient experience comments from a survey in primary care." *Journal of Medical Internet Research Medical Informatics.* 3 (2): e20. Doi: 10.2196/medinform.3783.

Marijan, Branka. 2020. "Armed Drones on the Canadian Military Horizon." *The Ploughshares Monitor,* 41.3 (Autumn). https://ploughshares.ca/pl_publications/armed-drones-on-the-canadian-military horizon/

McCammon, Sarah. 2017. "The Warfare May Be Remote but the Trauma is Real." *All Things Considered, NPR,* April 24, 2017. https://www.npr.org/2017/04/24/525413427/for-dronepilots-warfare-may-be remote-but-the-trauma-is-real/

Melzer, Nils. 2009. *Targeted Killing in International Law.* Oxford, United Kingdom: Oxford University Press.

Milne, Seumas. 2012. "The Drones of Washington Plough the Soil for Terror Reprisals." *The Age,* June 4, 2012: 9. ProQuest International Newsstream.

Monbiot, George. 2012. "For Whom Do We Cry?" *The Age*. December 24, 2012: 9. ProQuest International Newsstream.

Myre, Greg. 2023. "How Ukraine Created an 'Army of Drones' to Take on Russia." *NPR*, June 20, 2023. Accessed on June 21, 2023. https://www.npr.org/2023/06/20/1183050117/how-ukraine-created-an-army-of drones-to-take-on-russia/

National Archives of Australia. 2019. "Australia's Prime Ministers." Accessed April 15, 2019. http://primeministers.naa.gov.au/primeministers/

National Library of Australia. 2014. "History of Australia Newspapers." Accessed July 21, 2020. https://www.nla.gov.au/132anada132ia132-newspaper-plan/for researchers/history-of-australian-newspapers/

Neuendorf, Kimberly A. 2017. *The Content Analysis Guidebook*. 2nd edition. Thousand Oaks, California: SAGE Publications, Inc.

New America. 2019. "World of Drones." *New America*. Accessed March 27, 2019. https://www.newamerica.org/in-depth/world-of-drones/1-introduction-how-we becameworld-drones/

_____. 2021. "America's Counterterrorism Wars." *New America*. Accessed April 6, 2021. https://www.newamerica.org/in-depth/132anada132i-counterterrorism-wars/

The New York Times. 2019. "History." Accessed March 3, 2019. https://www.nytco.com/company/history/.

New Zealand Gazette. 2021. "New Zealand Gazette." *New Zealand Gazette.* Accessed April 20, 2021. https://gazette.govt.nz/

New Zealand Government. 2016. "The Press Goes to Press." *New Zealand Government, Ministry for Culture and Heritage*. December 20, 2016. Accessed December 9, 2020. https://nzhistory.govt.nz/page/lt-em-gt-press-lt-em-gt-goes-press.

_____. 2017. "Biographies – Premiers and Prime Ministers." *New Zealand Government, Ministry for Culture and Heritage.* November 1, 2017. Accessed March 3, 2019. https://nzhistory.govt.nz/politics/premiers-and-prime-ministers.

_____. 2020. "Samuel Revans Prints First Newspaper." *New Zealand Government, Ministry for Culture and Heritage*. October 20, 2020. Accessed April 20, 2021. https://nzhistory.govt.nz/page/133anada-revans-prints-first-newspaper

News Media Canada. 2020. "Daily Newspaper Circulation Data". Accessed April 1, 2021. https://nmcmic.ca/about-newspapers/circulation/daily-newspapers/

Nolen, Stephanie. 2012. "A Taboo Thought in Pakistan: What if Drones Work?" *The Globe and Mail*. November 17, 2012. ProQuest Canadian Newsstream.

Nye, Joseph S. 2008. "Smart Power and the 'War on Terror.'" *Asia-Pacific Review* 15 (1): 1 https://doi.org/10.1080/13439000802134092

_____. 2009a. "Get Smart: Combining Hard and Soft Power." *Foreign Affairs* 88 (4): 160 163.

_____. 2009b. "Smart Power." *New Perspectives Quarterly* 26 (2): 7–9. https://doi.org/10.1111/j.1540-5842.2009.01057.x/

______. 2011. *The Future of Power*. New York City, New York: Public Affairs.

Obama, Barack. 2010. "2010 White House Correspondents' Dinner." Filmed May 1, 2010, at the White House Correspondents' Association Annual Dinner. *C-Span*. https://www.c-span.org/video/?293275-1/2010-white-house-correspondents dinner

_____. 2020. *A Promised Land*. New York City, New York: Crown.

Oborne, Peter. 2011. "The Logic Behind bin Laden." *The Age*. May 9, 2011: 15. ProQuest International Newsstream.

Parks, Lisa. 2018. *Rethinking Media Coverage: Vertical Mediation on the War on Terror*. New York City, New York: Routledge.

Parliament of Canada. 2019. "Prime Ministers of Canada." 2019. https://lop.parl.ca/sites/ParlInfo/default/en_CA/People/primeMinisters/

Pew Research Center. 2013. "America's Global Image Remains More Positive than China's." https://www.pewglobal.org/2013/07/18/134anada134i-global-image remains-more-positive-than-chinas/

_____. 2015. "Public Continues to Back U.S. Drone Attacks: Afghanistan Update: Most Say U.S. Has Failed to Achieve Goals." http://assets.pewresearch.org/wp-content/uploads/sites/5/2015/05/5-28-15-Foreign-Policy-release.pdf

_____. 2019. "Global Indicators Database 2012-2014." Accessed May 5, 2019. https://www.pewglobal.org/database/indicator/52/survey/all/response/Disappr ve/

The Press. 2016. "Egged on with Egg on His Face." *The Press*. October 22, 2016, page C14. ProQuest International Newsstream.

ProQuest Canadian Newsstream. 2020. Accessed July 24 and 25, 2020. https://search-proquest-com.ezproxy.snhu.edu/canadiannews/fromDatabasesLayer?accountid=3783/

ProQuest International Newsstream. 2020. Accessed July 24, 25, 26, and 27, 2020. https://search-proquest-com.ezproxy.snhu.edu/internationalnews1/fromDatabasesLayer?accountid=378/

ProQuest U.S. Newsstream. 2020. Accessed July 25 and 26, 2020. https://search-proquest com.ezproxy.snhu.edu/usnews?accountid=3783/

Random. 2021. "Random Sequence Generator." *Random.org*. Accessed January 28, 2021. https://www.random.org/sequences/

Rinehart, Christine Sixta. 2018. *Drones and Targeted Killing in the Middle East and Africa: An Appraisal of American Counterterrorism Policies*. Lanham, Maryland: Lexington Books.

Rockwell, Geoffrey and Stefan Sinclair. 2016. *Hermeneutica: Computer-Assisted Interpretation in the Humanities.* Cambridge, Massachusetts: The MIT Press.

Roper Center. 2020. Approval from The Gallup Organization for 2008-2019. University of Minnesota subscription data sent via email.

Rudman, Brian. 2017. "*Trump Not First Bully President.*" *The New Zealand Herald*. February 8, 2017: A10. ProQuest International Newsstream.

Samios, Zoe. 2018. "ABCs: Newspaper Circulation Suffers Across the Board with Falls as Large as 16%." *Mumbrella*, August 17, 2018. https://mumbrella.com.au/abcs newspaper-circulation- suffers-across-the-board-with-falls-as-large-as-16-535665/

Sampsel, Laurie J. 2018. "Voyant Tools." *Music Reference Services Quarterly* 21 (3): 153 157. https://doi.org/10.1080/10588167.2018.1496754.

Savage, Michael. 2013. "UK Aircrew Flew 2,000 US Drone Missions." *The Times.* April 26, 2013: 9. ProQuest International Newsstream.

Sharkey, Noel. 2012. "International: Special Report: Analysis: Do We Want to Give Robot Killers to the World's Secret Services?" *The Guardian*. August 4, 2012: 26. ProQuest International Newsstream.

Shaw, Ian GR. 2017. "Robot Wars: US Empire and Geopolitics in the Robotic Age." *Security Dialogue* 48 (5) 451-470.

Sheets, Penelope, Charles M Rowling, and Timothy M Jones. 2015. "The View from above (and below): A Comparison of American, British, and Arab News Coverage of US Drones." *Media, War & Conflict* 8 (3): 289–311. https://doi.org/10.1177/1750635215593973.

Sinclair, Stéfan, and Geoffrey Rockwell. 2016. *Voyant Tools.* https://voyant-tools.org/docs/.

Singer, Peter W. 2009. *Wired for War: The Robotics Revolution and Conflict in the 21st Century.* New York City, New York: Penguin Books.

Snow, Shawn. 2020. "The US is Bombing the Hell Out of the Taliban as Munitions dropped reaches nearly 10-year High." *Military Times,* January 27, 2020. https://www.militarytimes.com/flashpoints/2020/01/27/the-us-is-bombing-the hell-out-of-the-taliban-as-munitions-dropped-reaches-nearly-10-year-high/

Seipel, Hubert. 2014. "Snowden Interview in English." *Norddeutscher Rundfunk.* January 26, 2014. http://www.ndr.de/ratgeber/netzwelt/snowden277_page 2.html

Springer, Paul J. 2013. *Military Robots and Drones: A Reference Handbook*. Santa Barbara, California: ABC-CLIO.

Statista. 2019. "UK Newspapers Ranked by Circulation 2019." *Statista*. June 24, 2019. https://www.statista.com/statistics/529060/uk-newspaper-market-by-circulation/

Sterman, David. 2022. "The State of America's Drone Wars in 2022." *New America.* Accessed June 21, 2023. https://www.newamerica.org/international security/blog/the-state-of-americas-drone-wars-in-2022/

Stratfor. 2019. "Visual Anthology: The U.S. Drone Fleet." *Stratfor.* Accessed September 8, 2019. https://worldview.stratfor.com/article/uav-visual-anthology-us-drone-fleetunmanned-aerial-systems-army-air-force-navy/

Thomas, Jason. 2018. "We Need to Understand Our Adversaries' Thinking." *The Australian*. March 14, 2018: 14.

US Air Force. 2007. RQ-4 Global Hawk. Accessed April 6, 2021. Photo by US Air Force and B. Zapka. *Wikimedia Commons.* https://commons.wikimedia.org/wiki/File:Global_Hawk_1.jpg

_____. 2008a. Photo of an MQ-1 Predator. Accessed April 6, 2021. Photo by US Air Force and Lt. Col. L. Pratt. *Wikimedia Commons.* https://commons.wikimedia.org/wiki/File:MQ-1_Predator,_armed_with_AGM-114_Hellfire_missiles.jpg

_____. 2008b. Photo of MQ-9/RQ-9 Reaper. Accessed April 6, 2021. Photo by US Air Force and Lt. Col. L. Pratt. *Wikimedia Commons.* https://commons.wikimedia.org/wiki/File:MQ-9_Reaper_UAV_(cropped).jpg

Voyant Tools. 2020. "Voyant Tools Help." Accessed July 31, 2020. https://voyant tools.org/docs/#!/guide

Wassef, Khaled. 2021. "Saudi Arabia Oil Facilities Targeted in Drone and Missile Attack by Iran-backed Houthis in Yemen." *CBS News.* March 8, 2021. https://www.cbsnews.com/news/137anad-arabia-drone-attack-oil-infrastructure rastanura-dhahran-houthis-yemen-iran/

Watson, Amy. 2021. "Newspapers in Canada – Statistics & Facts." *Statista.* March 5, 2021. https://www.statista.com/topics/2969/newspapers-in-canada/

Watson, Kathryn. 2019. "Trump Nixes Public Report on Civilians Killed by Drone Strikes." *CBS News.* March 6, 2019. https://www.cbsnews.com/news/trump-signs-executive-order-canceling-public-reports-on-civilian-drone-strike-deaths/

White, Brion. 2015. *Killing in Silence: Alternative and Mainstream Media Coverage of Drone Strikes.* Ph.D., United States – Ohio: Bowling Green State University. http://search.proquest.com/dissertations/docview/1700193386/abstract/8318C FD1004107PQ/1/

Whiton, Christian. 2013. *Smart Power: Between Diplomacy and War.* Washington, D.C.: Potomac Books.

Wilson, Ernest J. 2008. "Hard Power, Soft Power, Smart Power." *The ANNALS of the American Academy of Political and Social Science* 616 (1): 110–24. https://doi.org/10.1177/0002716207312618/

Wong-Diaz, Francisco. 2013. *Smart Power and U.S. National Strategy*. MacDill Air Force Base, Florida: Joint Special Operations University Press.

Woods, Chris, and Alice Ross. 2012. "Revealed: US and Britain Launched 1,200 Drone Strikes in Recent Wars." *The Bureau of Investigative Journalism*. December 4, 2012. https://www.thebureauinvestigates.com/stories/2012-12-04/revealed-us and-britain-launched-1-200-drone-strikes-in-recent-wars/

Worldpress.Org. 2019. "World Newspapers and Magazines." *Worldpress.Org*. Accessed March 27, 2019. https://www.worldpress.org/gateway.htm/

World Atlas. 2019. "The Most Popular Daily Newspapers in Canada." *World Atlas*. June 4, 2019. https://www.worldatlas.com/articles/the-most-popular-daily-newspapers-in- 138anada.html/

Wroe, David. 2019. "Lessons in Ethics for 'Killer Robots." *The Age*. March 1, 2019: 3. ProQuest International Newsstream.

Yin, Robert K. 2002. *Case Study Research: Design and Methods, 3rd Edition (Applied Social Research Methods, Vol. 5)*. 3rd edition. Beverly Hills, California: SAGE Publications, Inc.

Zenko, Micah and Sarah Kreps. 2014. *Limiting Armed Drone Proliferation*. Council on Foreign Relations, Council Special Report No. 69. June 2014.

Zucchino, David. 2019. "U.S. Drone Kills 30 Civilians, Afghans Say; Target Was ISIS." *The New York Times*. September 20, 2019: A6.

INDEX

ABOUT THE AUTHOR

Dr. Melissa Aho has earned an M.S. in Anthropology from the University of Wisconsin - Milwaukee, an M.L.I.S. from Dominican University, an M.A. in Art History from the University of St. Thomas, and an M.Ed. in Learning Technologies from the University of Minnesota - Twin Cities. Her Ph.D. in International Development is from the University of Southern Mississippi. She has participated in archaeological excavations in Belize, Colorado, Minnesota, and Wisconsin, researched mosque architecture in Guyana and Minnesota, and spent time in 16 countries. Dr. Aho has written books, book chapters, articles, and book reviews. She is a member of The Ninety-Nines - the International Organization of Women Pilots and The Explorers Club.